Contents

Acknowledgements

The authors are greatly indebted to Dr P. Armstrong (University of Western Australia), Dr D. J. Briggs and Dr N. R. J. Fieller (University of Sheffield), Dr K. McCloy (Satellite Mapping Unit, NSW Government, Sydney) and Dr A. R. Jones (Plymouth Polytechnic) for reading and commenting helpfully on the manuscript; particular thanks to Jenny Wyatt, who drew most of the Figures, and also to Brian Rogers and Sarah Cockerton for cartographic assistance; to Paul Webber, Ashley Simpson and Nick Trench for permission to use their unpublished field data; and to Gloria Day and Sue Cathcart for typing sections of the manuscript.

The following people and institutions have given us valuable information and assistance: Dr Susan Barker (Department of Environment, State Government of South Australia); Dr J. Carnahan (Australian National University); Dr John Cousens (Department of Forestry and Natural Resources, University of Edinburgh); CSIRO Division of Land Use Research; CSIRO Division of Soils; The Devon Trust for Nature Conservation; Mr Max Foale (University of Adelaide); Professor C. H. Gimingham (University of Aberdeen); Mr D. Glue (British Trust for Ornithology); Professor Sir Harry Godwin FRS (Cambridge); Professor V. H. Heywood (University of Reading); Mr J. A. Knight (State Geological Survey of Victoria); Dr Peter Laut (CSIRO); Mr David Moyle (Sturt CAE, Adelaide); Dr Oliver Rackham and the Cambridgeshire and Isle of Ely Naturalists Trust; Dr D. S. Ranwell (University of East Anglia); Dr T. Riley (Sheffield City Museum); Dr David Shimwell and the Kent Trust for Nature Conservation; Professor D. H. Spence (University of St Andrews); Mr A. Stead (Lancashire Naturalist's Trust) and the University of Arizona, Laboratory of Tree Ring Research.

We wish to thank the following who gave us permission to produce various figures, plates and tables used in this book: The Literary Editor of the late Sir Ronald Fisher FRS; Dr Frank Yates and the Longman Group Ltd, London, for permission to reprint Tables III, IV and XXXIII from their book *Statistical Tables for Biological, Agricultural and Medical Research*, 6th edition, 1974; The Division of Natural Resources, Mapping and Environment, The Government of Papua New Guinea.

All other sources are acknowledged in the text or in the appropriate caption.

Preface

Our aim in writing this book is to provide students and teachers with a simple introductory text which deals with practical aspects of ecology, environmental biology and biogeography, emphasizing actual field and classroom investigations. Basic concepts and methods of survey, mapping and aerial photography, data collection and data analysis are described and discussed, in order to encourage students to identify and tackle worthwhile projects.

The level at which this text is appropriate depends very much upon particular circumstances. The greater part lies within the scope of the sixth form and the first and second years of college, polytechnic and university courses in the British Isles and their equivalents overseas.

All students inevitably meet difficulties in the identification of plant and animal species, particularly when they venture into unfamiliar habitats and regions. This is often the cause of unnecessary alarm. Many ecological principles or problems may be illustrated by reference to familiar species and habitats, such as are found in urban environments, as well as those areas of semi-natural vegetation favoured for field courses.

At the end of each chapter we have included classroom and field exercises and projects, which may be tackled by the individual student or groups of students in a combined project. The examples illustrated are usually discussed in terms of one location or situation; however most projects may be adapted and modified very easily to suit particular circumstances and environments. The text and projects are fully cross-referenced in order to demonstrate the nature of relationships within ecology.

Simple quantitative techniques are introduced towards the end of the book, where they are used to solve ecological problems or to describe situations with more precision. All the calculations are easily carried out on a pocket calculator.

Most important of all, we hope that those reading and using this book will sense, and perhaps discover for themselves, the enthusiasm and enjoyment which we have found studying and teaching this subject with a practically orientated, problem-solving approach.

Safety in the field

All British college, school and university teachers and students must note the provisions of the *Health and Safety at Work Act, 1974*. All field work, even when conducted in local, well known areas, is potentially dangerous. Consequently precautions must be taken, no matter how trivial risks may appear. It is very easy to underestimate the problems and hazards in urban environments, especially when problems of pollution in aquatic ecosystems are being studied. Carry out a full reconnaissance of the survey area and ensure there are no problems of access or safety before beginning the field project.

Laboratory studies involving the use of acids, alkalis and scientific equipment are described in Chapter 14, which deals with pollution studies. These studies may only be attempted under supervision and in laboratories which are appropriately equipped.

You should obtain, study and act upon the recommendations in *Safety in Fieldwork* (Natural Environmental Research Council, no date) and *Safety on Mountains* (British Mountaineering Council, no date); addresses are in the appendix.

Everybody should be prepared to cope with sudden bad weather and a possible accident. Responsible members of the trip must be familiar with basic first aid objectives and procedures. The following equipment should be carried: waterproofs with hood, first aid kit, torch with batteries, whistle, emergency rations, including glucose sweets, map and compass, spare warm clothing, a tent or survival blanket. You should be able to survive at least a full day and night in adverse conditions. The standard SOS signal for torches or whistles is three short blasts followed by three long blasts and three short blasts, repeated.

Before setting out:

1 Obtain an up to date 'official' weather forecast and seek the advice of local experts if possible.

2 Obtain full details of any potential health problems of all members of the party. The information should be given to the member of staff in charge,

together with up to date contact addresses of family or friends. This information together with full details of the route and expected return time must be given to a responsible person at the base and a time stated at which the emergency services must be contacted.

Never, ever, carry out field work alone: a group of three leaves one person free to go for help and allows a second person to look after any injured person and speed the relocation of the injured person by the rescue services. Work in aquatic habitats, in caves or in quarries obviously involves special safety precautions and equipment, and may also involve various legal or insurance requirements.

It is always worth obtaining a tetanus injection to avoid the possibility of contracting the illness through scratched or grazed hands, while handling soils, from wire, or in farm yards.

When on surveys, never handle vertebrates unless you are under close supervision from an expert and are familiar with the risks involved. With dead animals or animal droppings, do not take the risk of infection, always wear protective gloves, clothing and face masks.

It is easy to ignore or belittle these concerns and precautions. Unfortunately one problem of increasing old age is that now we all *used to know* people who took too few precautions.

Part One
Basic Concepts and Principles

1 Tackling ecological problems

1.1 The purpose of this book

This book is an introduction to four important aspects of ecology, environmental biology and biogeography:

1 to sources of information;
2 to methods of data collection;
3 to techniques of survey and mapping;
4 to the analysis and interpretation of biological and environmental data using simple statistical methods.

Basic ecological survey and mapping techniques are described and in order to understand the data obtained, underlying ecological principles and concepts are introduced and discussed. A further aim of particular importance is to introduce students to the way in which the wide variety of approaches and techniques described here may be used to generate information of relevance to biological conservation and environmental management.

The text is appropriate to students at the very beginning of their studies and will help them tackle simple but worthwhile projects. There is a gradual progression through the book in the degree of complexity of the problems approached and the methods used to resolve them. Numerous case studies are presented, both as examples, and as models which students can use to establish their own investigations. Studies are described from many parts of the world – Africa, Australia, Europe and North America, and the project methods are also shown to be applicable, not only to semi-natural, but also to man-made environments because they are the types of habitats close to the homes of most students. Similarly, the book concentrates on species that will be familiar to many people, for example sea lavender, bracken, sheep and hedgehogs. At the end of most chapters, a series of questions, or individual and group projects for use in the class or in field work is presented, together with suggestions for further investigations.

There are several reasons for adopting this approach. Plants and animals are of considerable interest in their own right, as well as representing important resources in the landscape. Attempts to solve academic or applied problems of an ecological nature frequently necessitate the description and mapping of individual plants or animals and/or the communities they comprise. In many parts of the world, these activities are essential parts of basic resource data gathering programmes. Ecological knowledge is advancing rapidly. In a few years time, a number of present ideas and explanations will be found to be inadequate or incorrect. Consequently, it is important not only to have a sound grasp of the theoretical basis of the subject, but also to be familiar with the procedures for obtaining new information and advancing knowledge.

1.2   Spartina anglica – friend or foe?

The scope and ramifications of problems in ecology or biogeography are not always readily grasped. Something of the diversity of problems that occur can be illustrated by describing the history of a very common salt marsh grass called *Spartina anglica* (the common cord grass). The grass is illustrated in Plate 1.1. It can now be found along many of the estuarine shores of the British Isles (Figure 1.1).

Barnaby's Sands salt marsh SSSI

This account starts with a small coastal nature reserve – the Barnaby's Sands Marsh, which is typical of the type of small local reserve which might be used in student projects. However, this could just as easily start on the eastern or north-western coasts of North America, the major sea port of Southampton, or the temperate estuaries of western Europe, or Australia and New Zealand.

The coastal salt marsh at Barnaby's Sands occurs in the estuary of the River Wyre where it enters Morecambe Bay, opposite Fleetwood, Lancashire (Figure 1.2). The marsh is part of a nature reserve, which was initially classified as a Site of Special Scientific Interest (SSSI) under Section 23 of the United Kingdom National Parks and Access to the Countryside Act of 1949, which is now superseded by the 1981 Wildlife and Countryside Act.

A plan of the reserve is given in Figure 1.2. The distribution of salt marsh and associated habitats are shown on the vertical aerial photograph (Plate 1.2), which is seen to be an essential part of the biological data for the site. The salt marsh is well known to specialists and local naturalists for several reasons. It is particularly noteworthy for the richness and diversity of its salt marsh plants, especially the three species of sea lavender (*Limonium* spp) which display an attractive purple/lavender cover in late summer. The principal management problem at the marsh concerns the salt marsh grass *Spartina anglica*. The problem is seen in Plates 1.1 and 1.2. The grass is threatening to engulf, and eventually to suppress, many of the other attractive and ecologically important salt marsh species, including the lavenders. This

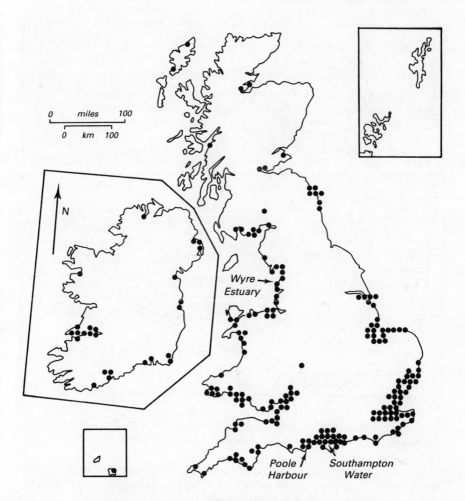

Figure 1.1 The distribution of the salt marsh grass *Spartina anglica* (Common cord grass) in Great Britain and Ireland in 1980; according to their presence or absence in 10 km × 10 km grid squares (By kind permission Biological Records Centre, Institute of Terrestrial Ecology.)

problem is simply the local manifestation of a long and surprisingly complex series of events, which have involved the impact of human activity on salt marsh ecosystems, accidental and deliberate transport of salt marsh grasses across oceans, hybridization, attempts to maintain deep-water navigation channels and coastal protection schemes in major estuaries.

First the local problem is explored in more detail before investigating the broader lessons to be learnt from the creation and spread of *Spartina* grass.

The first Ordnance Survey Map of the area in 1850 showed salt marsh to the east and north of Arm Hill, which was called Hame Hill at that time. Salt marsh was also mapped at Kneps Marsh to the south-east of the present reserve

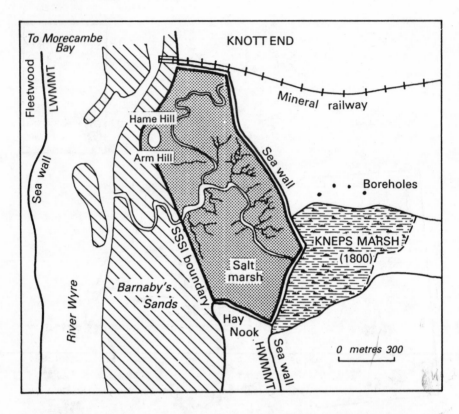

Figure 1.2 The Barnaby's Sands Nature Reserve (SSSI) in the Wyre Estuary, Lancashire (By kind permission of Lancashire Naturalists Trust, 1975/76.)

(Figure 1.2). The construction of a mineral railway with a pier and sea wall, in front of the underground salt extraction boreholes put down in the 1890s, caused the almost total loss of Kneps Marsh.

Until the 1930s the area between Arm Hill and the Hay Nook remained largely unvegetated as inter-tidal sandflats and mudflats. Since that time, an extensive salt marsh has developed in front of the sea wall which protects the now drained Kneps Marsh, and provides a counterpart of the long established salt marsh east of Arm Hill.

Three species of lavender are found in these marshes; the relatively common *Limonium humile* (lax-flowered sea lavender) and *Limonium vulgare* (common sea lavender), together with the much rarer *Limonium binervosum* (rock sea lavender). The variety of species and resultant rich purple colouring of the Barnaby's Sands salt marsh is in marked contrast to the great spreads of short grey-green turf which characterize many of the coastal marshes of Morecambe Bay and the Lake District to the north, as shown in Figure 1.1 and Plate 1.3. The Morecambe Bay salt marshes are dominated by sea poa, or the common salt marsh grass. These marshes are maintained as a very short turf by large numbers of grazing sheep and smaller numbers of grazing cattle. The resulting sea turf is extensively cut and transplanted for use as lawns, soccer pitches and in parks. Obviously this grass is able to tolerate a wide range of environmental conditions.

In contrast, Barnaby's Sands salt marsh has not been grazed in living memory and this has undoubtedly contributed to its richness and attractiveness. The different management and development of the two marshes, involving pumping, wall construction, drainage, grazing and exploitation have all combined to produce very different salt marshes on adjacent sections of coastline.

Incidentally, in spite of its common or coloquial name, sea poa is not a member of the *Poa* genus of grassses. The Latin binomial (name) used throughout the scientific world when referring to this grass is *Puccinellia maritima*. This demonstrates how misunderstandings arise from the use of the common name for a species and how these are avoided with the use of the Latin binomial.

The arrival of *Spartina anglica* in the Wyre Estuary

In 1938, local residents noticed the establishment of a yellow-green, erect grass on the inter-tidal mudflats of the Wyre Estuary. These were initially recorded as specimens of the fertile hybrid of *Spartina* known as *Spartina x townsendii*, now renamed *Spartina anglica*. The arrival of *Spartina anglica* at Barnaby's Sands marked one stage in the spread northwards of the species from its point of origin at Lymington, near Southampton, located in Figure 1.1. Figure 1.1 shows that the grass is now well established in salt marshes around the coast of Britain. Each cell on the map is a 10 × 10 km cell used by the Botanical Society of the British Isles and more recently by the Biological Records Centre as the basic mapping unit for biological surveys of the British Isles. Each dot indicates that at least one specimen in the grid square has been recorded and verified by reliable botanists (see Chapter 9).

A visit to Barnaby's Sands will soon show that *Spartina anglica* has now almost completely suppressed the *Salicornia* spp (the glassworts), which used to be the main species on the lower marsh. Field inspections or repeated detailed mapping over a period of years will also indicate that *Spartina* is actively invading other parts of the marsh.

It is difficult to envisage how to control the grass with minimal expense for salaries or special treatments. The local Naturalist's Trust, in whose charge the marsh is placed, is not a rich organization. Control of this pest species by introducing grazing animals is not practicable, since sheep or cows are likely to find the other marsh species more palatable, and hence reduce the marsh to the less attractive and less interesting state found on the heavily grazed marshes of Morecambe Bay immediately to the north. Attempts to control it by herbicide treatment, cutting or digging out have met with only limited success owing to the substantial costs of herbicides, frequent flooding, and the difficulties of access and working on the silty substrates of inter-tidal soils.

Plate 1.1   Tall, erect *Spartina anglica* colonizing open salt pans at Barnaby's Sands salt marsh (Photo: D. Gilbertson)

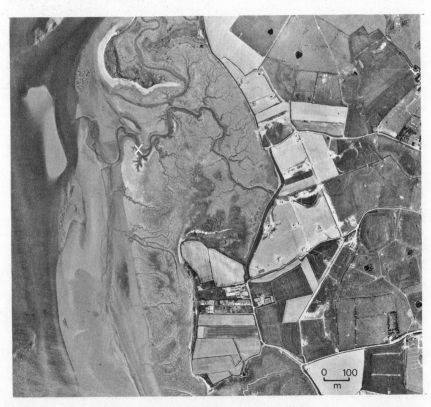

Plate 1.2   Vertical aerial photograph of Barnaby's Sands salt marsh, estuary of the River Wyre, Fleetwood, Lancashire in 1963; 'x' lies in areas of lighter tone associated with invading *Spartina anglica* (By kind permission of Huntings Aerial Photographs Ltd, and Lancashire County Council.)

Plate 1.3 Salt marsh comprising short turf of intensively grazed sea poa (*Puccinellia maritima*) at Cockerham Marsh, Morecambe Bay; 10 km north-east of Barnaby's Sands (Photo: D. Gilbertson)

Table 1.1 *Maximum recorded depth of silt accumulated and mean accretion rate for Spartina anglica marsh in different parts of the world* (After Ranwell (1967), by kind permission Blackwell Scientific Publications Ltd.)

Country	Site	Tidal range (m)	Depth of silt accumulated by S. anglica (m)	No. of years	Mean annual accretion (cm)
New Zealand	Waihopai estuary	2.7	+0.61	30	2
Tasmania	Tamar estuary	3.7	+0.46	16	3
British Isles	Poole Harbour	1.8	+0.91	50	2
British Isles	Bridgwater Bay	12.0	+1.83	37	5
Netherlands	Sloedam	3.8	+1.80	22	8

The following extreme accretion rates have been recorded for short periods only in young actively growing marshes:

British Isles	Bridgwater Bay	12.0	+0.17	1	17
Netherlands	Sloedam	3.8	+1.00	5	20

Table 1.2   *Dates of introduction and areas of Spartina anglica salt marsh in different parts of the world* (After Ranwell (1967), by kind permission Blackwell Scientific Publications Ltd.)

Country	Date	Area in hectares	Country	Date	Area in hectares
Great Britain	1870	12,000	France	1930	4000–8000
Ireland	1925	200–400	Australia	1930	10–20
Denmark	1931	500	Tasmania	1927	20–40
Germany	1927	400–800	New Zealand	1913	20–40
Netherlands	1924	4000–5800	United States	1960	less than 1
Total		21,000–27,700 ha			

Spartina grass and the protection of coasts and deep-water navigation channels

Elsewhere, the salt marsh grass attracted the attention of civil engineers, geologists and Dock and Harbour Authorities. These people were concerned with problems of coastal erosion and with siltation in estuaries. Their interest lay not only in the rapidity with which *Spartina anglica* could colonize and spread along coastal mudflats, but also in its remarkable ability to trap and grow through silt brought in by tides. This capacity is indicated in Table 1.1. The data show that marsh levels are raised by silt accretion far more rapidly with *Spartina anglica* than occurs with most other salt marsh species. Potentially this property can assist by strengthening sea defence works (causing the waves to break further seawards than would otherwise occur) and in salt marsh reclamation projects. The consequent reduction of silt circulating in estuaries was also believed to reduce the frequency of, and necessity for dredging to maintain sufficient draft for the increasingly large ships entering ports and harbours located in tidal estuaries. An added bonus was that in areas lacking the grazing-sensitive salt marsh species, the grass was found to be palatable to certain breeds of cattle and sheep.

In consequence, a worldwide demand for this particular *Spartina* grass arose for coastal engineering, protection and navigation schemes. The response in the source country, the United Kingdom, was the establishment of two commercial nurseries. These were on the inter-tidal mudflats of Poole Harbour, Dorset, in southern England and on the marshes of the Essex coast (Figure 1.1). From these nurseries the grass was sold and distributed around the British Isles, or exported abroad. The records of the two main distributors were located and examined by Dr D. S. Ranwell and combined with his field and literature surveys to produce a picture of how far the grass had been exported.

The following estuaries in the British Isles received substantial numbers of

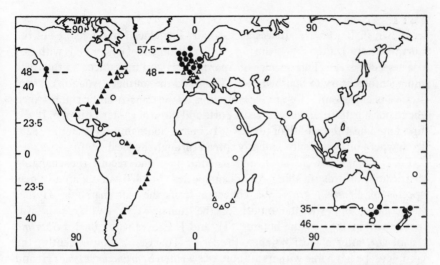

Figure 1.3 The location of successful establishment (closed circles) and unsuccessful transplantations (open circles) of *Spartina anglica* (After Ranwell (1967, 1972.) The distribution of *Spartina alternifolia* (closed triangles) and *Spartina maritima* (open triangles) – (After Mobberley (1956).)

transplants; the Thames, Blackwater, Humber, Belfast Lough, Severn and Exe. From these sites the grass has been further transplanted, or spread naturally into new locations, bringing benefits or problems depending upon the point of view. Dr Ranwell's estimates of the areas of *Spartina anglica* in different countries is given in Table 1.2.

There are unexpected scientific benefits in this large and complex set of transplants. Figure 1.3 indicates those locations in which the environmental conditions appear to have been beyond the tolerance range of the transplants. Examination of such maps, which reveal the geographical spread of a species, can in certain circumstances provide important information on the character and operation of those limiting factors controlling the abundance and distribution of organisms. Inspection of this map suggests that light or temperature may be important.

The history of *Spartina anglica* and trans-Atlantic trade

The history of *Spartina anglica* begins with two other members of the genus *Spartina* and an infertile hybrid that developed between them. *Spartina maritima* is nowadays an uncommon, even rare, salt marsh grass on the western coasts of Europe, with a major colony occurring at the head of the Adriatic and other colonies growing on the coasts of southern and south-western Africa.

The second species concerned is *Spartina alternifolia*. This species is widespread in tidal flats along the eastern shores of North and South America as shown in Figure 1.3. It was accidentally introduced to the shores of the north-western United States during the establishment of oyster beds with east coast seed oysters. This species of *Spartina* was also introduced in the early nineteenth century to Southampton Water and the southern Atlantic coast of France near Bayonne (Figure 1.3), by ships on the trans-Atlantic run between the major north American east coast ports and those of western Europe. It was first noted near the mouth of the River Itchen in Southampton Water in 1829, from where this grass subsequently spread through the local salt marshes.

Local botanists were interested to find, plot and record the changing distribution of this invading American species. In 1870, an apparently new species of *Spartina* grass was collected from the salt marsh at Hythe, immediately opposite the mouth of the Itchen. The new *Spartina* was 'described' in 1881 by the botanists H. and J. Groves who called it *Spartina townsendii*, after a local botanist Frederick Townsend. Consequently, the complete species name with its authors was written *Spartina townsendii* H. and J. Groves 1881.

Later studies have shown that the specimens described by Groves in 1881 were male-sterile hybrids which had resulted from a cross between the rare, but native *Spartina maritima*, and the accidentally introduced American species *Spartina alternifolia*. Another essentially similar hybrid developed independently on the French coast. To indicate the hybrid origin of the grass the name came to be written as *Spartina x townsendii*.

The story now becomes more complex. A fertile hybrid of this grass then evolved, with the earliest specimens being collected in 1893 from Lymington Water, a small estuary 20 km to the west of Southampton Water. By 1907 this new fertile hybrid had spread through salt marshes from east Dorset to the Suffolk coast. Recently, detailed studies by Dr C. E. Hubbard have shown that this distinct new form of *Spartina* should be designated as a separate species. It developed in England, hence *Spartina anglica* (Hubbard). The brackets around Hubbard indicate that Dr Hubbard was responsible for the re-designation of the species.

Spartina anglica is the grass that has been so widely transplanted or spread naturally throughout so much of the world. It has largely replaced its parents – *Spartina alternifolia*, *S. maritima* and *S. x townsendii* which have become rare on salt marshes. The story continues with further hybridization taking place, with infertile hybrids developing as a result of further crosses between *Spartina anglica* and *Spartina x townsendii*.

Most of the maps, tables, reports and charts published before 1970 which mention *Spartina x townsendii* and fertile *Spartina x townsendii* probably refer to *S. anglica*. It also seems likely that some of the early unsuccessful transplants of *Spartina* grass into regions where a good result might have been expected (see Figure 1.3), may have been the result of the inadvertent transplant of the infertile hybrids of *Spartina anglica* with *Spartina x townsendii* and not of the fertile *Spartina anglica*, which would have been termed fertile *Spartina x townsendii* at the time.

A diverse set of lessons

There are many lessons to be learnt from this history of a new species, starting with simple observations of a new grass arriving on a Lancastrian salt marsh. For example, the large-scale embankment and marsh reclamation at Kneps Marsh have been associated with the generation of an ecologically interesting and very attractive salt marsh. The very intensive grazing of the Morecambe Bay marshes also has its benefits producing the famous Cumberland sea turf, although from a biological point of view the resulting marsh is much less diverse and less attractive than the ungrazed swards of sea lavenders to the south (Plates 1.1–1.3).

The accidental and deliberate transhipment of species has had totally unexpected biological, geographical and geological consequences; in this case the creation of a new species and new hybrids with remarkable properties. In some respects, great advantages have accrued from these events, in others there have been unfortunate biological consequences which are proving very difficult to manage, or reverse, given the resources available.

The importance of precise identification and taxonomy are self-evident for both the specialist investigating limiting factors and the applied scientist seeking to control erosion or deposition by biological means. The study has involved the skills of the botanist, coastal geologist, geographer, engineer and surveyor, as well as causing the research workers to inspect aerial photographs, old maps and charts, nineteenth-century natural history journals and sales dockets!

This case study also illustrates the overlaps and relationships between subject areas, which are now given the various titles of ecology, environmental biology, biogeography and environmental science. All these disciplines are concerned with field surveys and data collection for solving environmental problems, albeit from different starting points. It is these common threads of survey, mapping, data analysis and interpretation that form the basis of this book.

1.3 The problem solving approach

Time, money, resources and energy are always at a premium. Consequently, the collection, analysis and presentation of biological or environmental data to resolve problems, such as the *Spartina* problem, must be approached in a logical and systematic fashion. This necessitates a scientific framework for the inquiry. If we wish to describe and explain the distribution of plants and animals, it is important to have a clear objective in mind. For example, it may be important to establish the likely impact of *Spartina* grass invading estuarine salt marshes upon both the pattern and rate of sediment accretion. Alternatively, the subject of concern might be the likely impact of the grass, both directly and indirectly by altering the light environment and sedimentation rate, on other salt marsh species. These questions are unlikely to be answered without a very clear research design. It is not always justifiable to want to survey plant and animal resources just because they are there. Other considerations may be more important and relevant.

Hypotheses

Clear reasoning and the precise statement of objectives are aided by the formulation of hypotheses. Hypotheses are ideas of what we might expect to discover in a particular situation. They are possible explanations of observations and problems. Hypotheses are tested to establish their adequacy as explanations of the phenomena and effects observed. They are derived from previous knowledge, current reading, intuition, and familiarity with the available techniques of study. For example, from previous reading it is possible to hypothesize that the rate of sedimentation on a *Spartina anglica* marsh will be determined primarily by the height of the *Spartina* sward. Second, we might hypothesize that other salt marsh species might increase in frequency, size or vigour, and the overall sedimentation rate decline, if the *Spartina* were controlled by a mechanized cutting programme.

These ideas may be tested in various ways. First, the careful use of old documents, maps, aerial photographs, hydrographic charts, paintings or pictures, will enable us to establish the patterns of change that have occurred in the recent past. Knowledge of the period before the arrival of *Spartina anglica* is very interesting, since it provides 'baseline' data which may enable us to identify in more detail the nature of the impact of *Spartina*. This type of study uses the 'pre-*Spartina*' period as a type of 'experimental' control and is often known as a 'before-and-after' experiment. The extent to which the results will be interpretable in terms of the original hypotheses will also depend upon the extent to which other environmental changes were occurring during this period and on the effects of human activity in the critical phases before, during and after colonization.

A second approach would be to re-map the marsh over a period of time and establish the pattern of species spread and change – before, during and after the period of *Spartina* invasion, colonization, or cutting. Precise topographic survey involving mapping and levelling (Part Three), or by using height marker pins inserted into the marsh, should enable identification of the changing height of the marsh surface with accuracy. If the results are to be of lasting value, then the data should be organized and analysed in a logical and statistically reliable manner (Part Four). The more carefully a particular problem and study method are thought out in advance, the more likely it is that a sound hypothesis will be generated, resulting in a satisfactory conclusion.

Rejected hypotheses

If a hypothesis is shown not to be an adequate explanation of the effect observed, it is not a disaster. Studies need to be organized so that a hypothesis can be shown to be inadequate or incorrect. It is important to show that a particular factor or process has not caused the observed effect. A rejected hypothesis means that ideas may have to be reformulated and a new hypothesis devised, perhaps tackling the problem from a different standpoint, or changing the assumptions about what is cause and what is effect. In the *Spartina* example, our first study might suggest that it is the mean height of the marsh surface, rather than the height or density of the *Spartina* sward that is the principal determinant of sediment accretion rates in the area.

The researcher may go through several successive hypotheses, each time changing the ideas, before he or she decides they have a satisfactory explanation and solution to a problem.

Multiple working hypotheses
Many field scientists have adopted Chamberlin's, method of multiple working hypotheses to help their studies (Chamberlin, 1965)*. The approach is to consider as many likely explanations as possible, with investigations then being designed to provide evidence that will lead to the rejection of as many of these multiple hypotheses as possible. This leaves the remaining one or two working hypotheses as the subject for further experiments or surveys. Frequently the scientific principle called 'Occam's razor' (from William of Occam, *c.* 1300–49) is applied. Briefly this indicates that the simplest, most straight-forward explanation should suffice, until there is good reason to suspect that a more complex explanation is required.

Collection of baseline data

Unfortunately in many parts of the world, it is not easy to devise hypotheses to explain the distribution, abundance or character of plant or animal resources. Fundamental knowledge of the vegetation and its associated animal life may be absent, or at best generalized. As a result, one large and essential area of research is concerned with providing a fund of baseline data, which may be used in the future for hypothesis generation and testing. In the meantime, this type of data-gathering approach forms an integral part of many general resource inventory surveys in developing and developed countries. These data may provide the basis for biological management and conservation programmes.

Difficulties

Several potential problems face the student. The first concerns identification of the plants and animals found. In this book readily identifiable or well known

---

*Full references quoted in text are contained in the References beginning on p. 301.

species are studied. Important study methods are also introduced for which it is not necessary to identify the species present. In situations where reliable identification is essential, samples may be labelled A,B,C, etc., and taken to the local expert or museum for identification. This is the procedure which field workers would adopt in a region which is unfamiliar to them. In general, and particularly where rarities are known to occur, specimens should not be collected; instead good quality colour photographs should be taken.

The discussions of faunal surveys have avoided techniques which include unnecessary and often unjustified trapping, killing and preserving. Biology is a life science. Instead we have concentrated on immediately identifiable species which are left alive at the end of the project.

The second difficulty will concern handling data. Quantification is usually essential if interesting and effective projects are to be developed. The statistical methods described here form a natural progression in difficulty and all may be applied by students who have taken basic mathematics courses. Only the ability to use a pocket calculator is required.

2 Fundamental concepts

2.1 Introduction

A number of concepts are central to ecology and biogeography and provide a framework for the disciplines. The first and most important is the ecosystem concept (Figure 2.1), which provides a major functioning model for all study of the interactions between living organisms and their environment. Second is the related concept of the plant and animal community. Next are the ideas of limiting factors, competition, successional processes and population dynamics. Together, these processes and controls determine much of the productivity, distribution and abundance of organisms. We are omitting discussion of internal biological controls such as physiology, state of health, genetic factors, species strategies and evolutionary concepts as these require specialist study.

2.2 The ecosystem

The ecosystem is an identifiable unit consisting of living organisms (the biotic component) and their physical and chemical environment (the abiotic component). Ecosystems are identified by their structure and function. By structure, we mean their physical components which we can see, feel and relate to with all our senses. Thus, all species of plants and animals are part of an ecosystem structure, as are the non-living rocks and sediments which underlie all ecosystems and the climatic factors which prevail upon them. By function we mean the interactions between the component parts of ecosystems. The functioning of ecosystems is concerned with the flow of energy and the cycling of nutrients through the system. The ultimate driving force for these processes within the ecosystem is solar radiation, which is converted to plant tissue in photosynthesis. In the conversion, exchanges of materials also occur. Soil nutrients and water are required and have to be circulated for the process to work effectively.

The functioning relationships between organisms are expressed in the idea of trophic levels. All plants and other organisms which obtain their energy

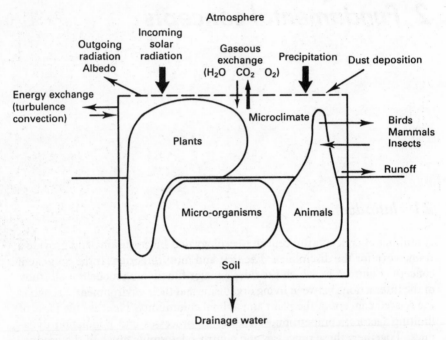

Figure 2.1   Schematic representation of an ecosystem or a biogeocenose (framed) in a state of exchange with the environment. If it is possible to recognize a distinct boundary to the ecosystem, for example, the edge of a woodland, then the area enclosed by the frame would constitute a biogeocenose (Redrawn from Walter (1973); by kind permission of Springer-Verlag, Heidelberg.)

from solar radiation through photosynthesis are classified as primary producers or the first trophic level. Animals which feed on these are known as herbivores or the second trophic level. They in turn are preyed upon by carnivores at the third trophic level. A fourth level often also exists comprising species such as omnivores which feed on all lower trophic levels.

The place in which a species lives is referred to as its habitat. Its position can be defined, not only in terms of spatial location, but also in terms of its role and functional significance in the ecosystem. Here we use the term ecological niche; the terms parasite, herbivore, carnivore or omnivore all describe a species' ecological role or niche.

The term ecosystem was introduced to Britain by the botanist Sir Arthur Tansley in 1935. Tansley appears to have thought of ecosystems as distinct vegetation units such as a salt marsh, mixed or oak wood, alpine pasture or a freshwater lake (Plate 2.1). Together with their associated animal populations and habitat conditions – soil, topography, microclimate – they constitute well-defined units within which distinct functional relationships can be identified.

Plate 2.1 Plant zonation in Loch-a-Raonabuilg, Isle of Colonsay, Inner Hebrides, Scotland (1) a band of white water lily *Nymphaea alba*, surrounds a central oval of deep water (probably with submerged plants, about 0.25 ha in extent); (2) zone of sedges; (3) zone of mixed vegetation at the edges, including sedges, grasses, water mint, bog myrtle, heathers and *Sphagnum* mosses – see also Figures 5.6 and 8.3. (Photo: D. Gilbertson)

2.3 The plant community

Vegetation is the most obvious external feature of many terrestrial ecosystems. It provides the food supply for all other organisms and also acts as a habitat within which most animals grow, live and reproduce. As a result, we tend to use vegetation units to define the area or extent of an ecosystem. If we stand on a hilltop and survey a landscape dominated by vegetation, the major units which we will recognize will often be those formed by the plant communities present. Major distinctions can be made on the basis of the physiognomy or the growth form of the vegetation; for example recognizing woodland or scrub as opposed to grassland (Section 4.3). More subtle differences in the landscape will be evident in variations in colour between areas of vegetation which have the same physiognomy. Here the recognition of plant communities requires the examination of its species composition (Section 5.1). However defined, the plant community is represented by a particular grouping or assemblage of species and this corresponds to one of the most important spatial units of ecosystem study for the ecologist or biogeographer.

Where natural or semi-natural vegetation has been substantially modified by man, plant communities in the sense discussed above, may have only limited presence. However, ecosystems can still be delimited, usually as different types of land use. Such land use units are often much easier to recognize than plant communities, since man creates very clear cut boundaries to different types of land use.

2.4   Limiting factors

The productivity, abundance and distribution of plants and animals are controlled by five major groups of factors:

1   Climatic factors, e.g. solar radiation, temperature, humidity, precipitation, wind, exposure;
2   Edaphic factors, e.g. soil physical properties, moisture and water relationships, soil chemistry and nutrient status;
3   Biotic factors, e.g. the effects of grazing, dunging and trampling by animals; and burning, cutting, harvesting or transportation by man;
4   Geological and evolutionary factors, e.g. global tectonics and continental drift, long-term climatic change;
5   Internal biological factors, e.g. species strategies and competitive mechanisms, successional processes, health, age, condition, genetically-based variations in tolerance or productivity.

Some or all of these factors determine the distribution of a species across the surface of the earth. Many species have very wide distributions in terms of their tolerance of climatic, edaphic or biotic controls – such species are said to be eurytypic in their distribution. Others are very narrow in their range and may only be found in very localized areas with the particularly favourable conditions in which they thrive. These are said to be stenotypic. If one factor, such as soil moisture, is taken, then plotting the response of a species along this environmental gradient in terms of its productivity or abundance will usually result in a normal or gaussian curve (Figure 2.2). The species of plant or animal will be most abundant, or the individual most productive where soil moisture conditions are at the optimum. The relative abundance tails off in the areas of increasing stress on either side of the optimum value. The points in Figure 2.2 where the species can no longer survive are termed the upper and lower limits of tolerance.

Similar curves can be drawn for the response of a species to any factor within the five groups outlined above. For a species to survive the environmental conditions at a point on the earth's surface, those conditions must be within the tolerance range of the species with regard to all factors at that site. If one factor is beyond the limit of tolerance of that species, then the species will not be able to survive. That factor is known as the master limiting factor. This is an important concept because often only the one factor needs to be limiting in order to prevent the growth and establishment of a species.

In many cases, the shortage preventing growth in plants may not be of a

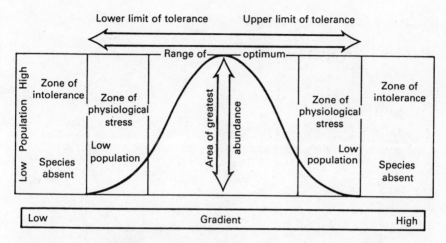

Figure 2.2 The normal curve of species distribution, showing the likely pattern of abundance or productivity of a species as an environmental variable changes (Redrawn from Cox and Moore (1985); by kind permission of Blackwell Scientific Publishers Ltd.)

macro-nutrient such as nitrogen or phosphorous, which is needed in fairly large quantities for satisfactory plant growth, but in substances such as boron or molybdenum which are essential micro-nutrients needed in minute quantities and only present in small amounts in the soil. These concepts are incorporated in Leibig's, *Law of the Minimum*, named after the nineteenth-century scientist, who stated that 'the growth of a plant is dependent on the amount of foodstuff which is presented to it in minimum quantity'. The identification and correction of deficiency in micro-nutrients to plants and animals are among the most important achievements of the present century and are responsible for bringing vast areas of land into more productive and economically more useful condition.

In some cases, however, this theoretical view of limiting factors may be too simplified, because factor compensation may occur. This happens when ideal conditions in terms of one factor for a species may compensate for conditions beyond the species' normal tolerance of another factor. This introduces the important principle, that while the distribution of some organisms may be studied in relation to a single environmental factor, a complete explanation will probably only result from a study of several factors and their effects in combination.

Laboratory and field investigations of environmental tolerance

The precise demonstration of the significance of interactions between limiting factors may require laboratory investigations. Figure 2.3 shows the results of experiments on the tolerance of the American lobster in relation to temperature, salinity and oxygen content in its coastal environment.

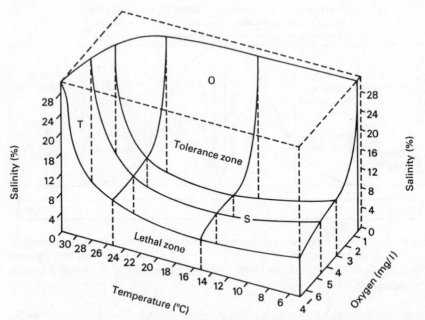

Figure 2.3 The tolerance limits of the American lobster established by laboratory study (Redrawn from McLeese (1956); by kind permission of the Fisheries Research Board of Canada.)

The lobster will survive up to a temperature of 28°C, as long as the salinity of the water is more than 14 parts per thousand and the oxygen content is more than 3 mg/l.

Valuable as these experiments are, it may often be necessary to question the validity of the results. Few laboratory experiments can adequately replicate the complex of interacting factors operating in the field.

Climatic controls

Field investigations may be less precise but are much closer to reality. A classic study of climatic limiting factors operating on holly (*Ilex aquifolium*), ivy (*Hedera helix*) and mistletoe (*Viscum album*) at the northern margins or their European distributions was published by Iversen (1944). The distribution of the holly through Europe and North Africa is illustrated in Figure 2.4. The north-eastern limit runs through Denmark and during the very cold winters of 1939–42, Iversen made observations on the extent to which the holly tree was damaged or even killed (Figure 2.5). The performance of the tree within a 20 km radius and 40 m height range of meteorological conditions was monitored. The results of his field surveys of tree distribution and health were plotted on a scattergram showing the mean temperatures of the coldest and the warmest months (e.g. Table 2.1; Figure 2.6). These values provide a measure of the

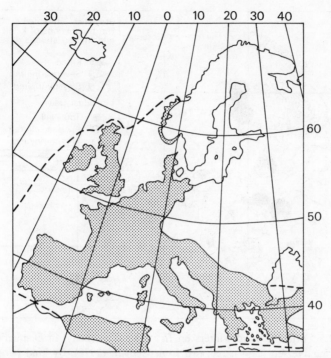

Figure 2.4 The distribution of the holly (*Ilex aquifolium*) in Europe and North Africa (Data from Iversen (1944); redrawn from Godwin (1975); by kind permission of Cambridge University Press.)

extremes of temperature experienced, the length of the growth period and the relative oceanicity or continentality of the site concerned. It is clear from this simple ordination diagram that holly is intolerable of winter cold but does not demand high summer temperatures; it requires an oceanic climate.

Figure 2.6 shows that higher summer temperatures – above 15°C in the warmest month – counterbalance to only a very minor degree, the impact of increasingly cold winters. Compare this observation with the plot of similar climatic data for the mistletoe (*Viscum album*) presented in the exercise at the end of this chapter. Mistletoe is a species better adapted to continental conditions.

The climatic control observed here does not have to occur every year to effectively limit distribution. Occasional determinations of climatic factors at the edge of distributions may well miss the onset of those conditions that most influence distributions. The margins of a species' range are best considered as a zone, into which the plant or animal is constantly trying to spread, establish and successfully reproduce itself. However, at various times, with varying degrees of magnitude, the master limiting factor operates. The boundaries of a distribution consequently consist of a zone, known as an ecotone, across which the individuals of a species become progressively fewer, less productive and

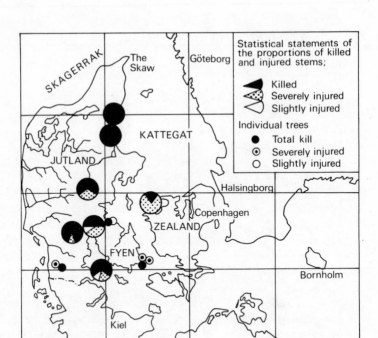

Figure 2.5 Effect of hard winters on *Ilex aquifolium* (holly) in Denmark 1939–42 (Data from Iversen (1944); redrawn from Godwin (1975); by kind permission of Cambridge University Press.)

perhaps smaller, until no more specimens are seen. This zone may measure a few metres or many kilometres, depending on the tolerance range of the species and the severity of the environmental gradient.

Edaphic controls

Figure 2.7 illustrates variation in tolerance ranges and the interaction of limiting factors due to soil conditions. The presence of four woodland ground flora species in Indiana, USA have been plotted in relation to the calcium and organic matter content of their soils. Each small circle represents a soil sample taken from the woods and the large circles indicate the presence of the species where the soil sample was taken. *Viola striata* (Cream violet) clearly has the widest range of tolerance to variation in both calcium and organic matter. In contrast, *Prunus serotina* (Wild black cherry) is only found where calcium levels are between 2 and 4 per cent, although over a range of organic matter content. *Sanguinaria canadensis* (Bloodroot) occurs only where soil organic matter is around 5 per cent but over a range of calcium values. *Viola eriocarpa* (Common yellow violet) is intermediate in its requirements. Each species is seen to respond differently to the combination of these two edaphic factors.

Biological modification of environmental conditions

Plants and animals do more than simply respond to environmental conditions. To a notable degree, they also modify their own environment. Temperatures inside a boreal pine forest in winter may be two or three degrees higher than those on surrounding snow-covered grassland. Vegetation creates its own micro-climate. The trees re-radiate heat from their trunks and crowns and trap pockets of air, reducing the effects of wind and exposure. In misty areas, condensation and fog drip from trees may greatly increase the total precipitation reaching the ground, while the same crowns may reduce the velocity with which moisture reaches the ground. Thus, before any climatic fluctuation can cause a change in the distribution of organisms, the magnitude of the change must exceed an ecological threshold to overcome the inertia or resistance to change, which exists because of the ability of the species and communities of the ecosystem to alter their micro-climate and environment.

Similar relationships exist between organisms and soils. When given a slight helping hand at initial planting, more calcium-demanding tree species such as beech (*Fagus sylvatica*) or birch (*Betula* spp.) may have the effect of increasing the amount of calcium in circulation between tree and soil, leading in the longer term to a progressive upgrading of soil status. On soils prone to leaching or in damp environments, the replacement of calcium-demanding deciduous trees with relatively undemanding shrubs such as ling (*Calluna vulgaris*) may lead to less recycling of calcium, which is then progressively lost through soil leaching. Eventually, this may cause serious soil deterioration.

2.5 Competition

The potential survival of particular species of plants and animals in an area is not controlled entirely by environmental factors. Competition between species (inter-specific competition) is very important. Usually the environmental conditions at a point on the earth's surface are suitable for more than one species to survive. The species which eventually occupies that point will have undergone competitive interaction (i.e. competed with, suppressed or eliminated other species). For example, in plant species, factors such as reproduction methods, growth rate, growth form and phenology (the timing of growth, flowering and fruiting) may influence the final outcome. Such competitive mechanisms are many and diverse. In plants, shade-casting properties are very important. The concept of species strategies is also relevant here in that, through evolution, species develop adaptations to increase their chances of survival against other rivals. Individuals of the same

Table 2.1   *Meteorological data for stations, within, on the border of, and beyond the range of Viscum album (Mistletoe) in Denmark* (Data from Iversen (1944), with kind permission, Cambridge University Press.)

Meteorological Station located with:	Meteorological Station Recording	
	Mean Temp. Warmest month °C	Mean Temp. Coldest month °C
Viscum album	16.5	5.4
	16.8	5.0
Growing within station area	16.6	3.7
(20 km radius)	17.7	2.4
(40 m height band)	18.3	2.3
Plot as a closed circle	17.5	−0.2
	17.9	−0.4
	18.8	−0.1
	19.1	0.4
	17.7	−0.2
	18.5	−1.0
	18.3	−2.3
	18.2	−5.5
	19.5	−7.0
	19.5	−1.8
	20.8	−5.8
	21.5	1.3
Viscum album Meteorological station on the edge	+15.8	+3.8
of the distribution of *Viscum*	15.9	3.2
	16.8	1.5
Plot as an open circle	17.2	0.9
	17.3	0.0
	16.7	−0.6
	17.0	−1.2
	16.9	−1.8
	16.6	−2.6
	16.9	−4.2
	17.4	−4.3
	17.8	−4.8
	18.2	−6.8
	20.8	−7.8

Table 2.1(*cont.*)

Meteorological Station located with:	Meterological Station Recording	
	Mean Temp. Warmest month °C	Mean Temp. Coldest month °C
Viscum album		
Absent from meteorological station	15.2	5.8
area	13.5	4.3
	14.8	2.8
Plot as a triangle	14.2	1.2
	14.0	0.1
	16.4	1.2
	0.5	16.2
	15.5	0.2
	15.9	−0.1
	15.5	−0.5
	16.5	−0.2
	16.4	−0.2
	16.2	−1.0
	16.8	−1.2
	16.4	−2.3
	12.8	−1.8
	15.5	−3.6
	16.6	−4.4
	17.0	−5.9
	17.0	−8.6
	15.8	−9.8
	19.2	−9.5

species also compete with each other in a similar manner. This is known as intra-specific competition.

2.6 Ecosystem change through time

Populations of plants and animals, plant and animal communities and ecosystems are subject to change through time. Soils similarly are affected, often being either improved or downgraded, depending upon particular circumstances, and hence they in turn can affect the flora and fauna of an area. These changes are usually brought about slowly by the interaction of many factors – natural biological processes and fluctuations, seasonal and longer term climatic fluctuations, geomorphic processes, human activity over long or short periods or a new factor starting to exceed a critical ecological threshold.

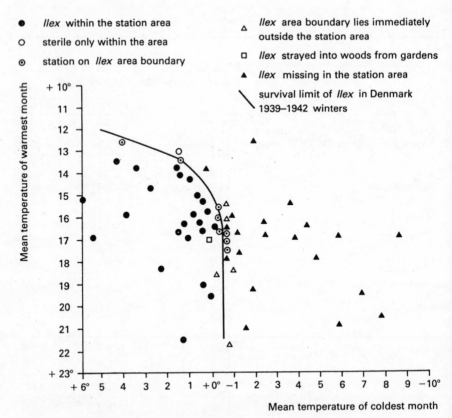

Figure 2.6 Scattergram of holly (*Ilex aquifolium*) performance in Denmark, in relation to mean performances of the warmest and coldest months in the period 1939–1942 (Data from Iversen (1944); redrawn from Godwin (1975); by kind permission of Cambridge University Press.)

Here we consider three important and interrelated concepts; population dynamics, succession and climax.

Population dynamics

The population of a species at any given point in time and space reflects the interplay between birth rate, death (mortality) rate and gains or losses due to migration. The population will have an age structure which represents the frequency of individuals of different ages. Two types of age structure commonly occur.

1   An even-age structure where most organisms are of the same age. This is characteristic of short-lived organisms, where for example, life begins in spring, individuals then develop, reproduce and die more or less together in autumn or early winter. Many forestry plantations also have this structure,

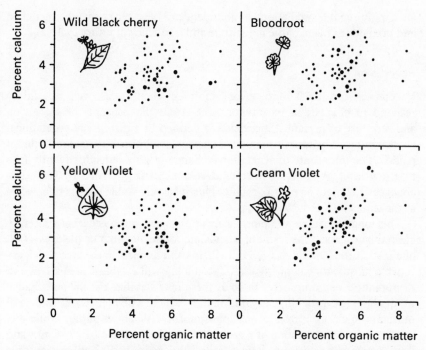

Figure 2.7 Distribution of four Indiana woodland ground flora species, in relation to percentage of soil calcium and organic matter; small dots show samples, large dots indicate the species presence (Redrawn from Beals and Cope (1964); by kind permission of Duke University Press and the Ecological Society of America.)

reflecting initial dense planting of large areas at the same point in time and subsequent suppression of any self-sown seedlings. Some 'natural' woodlands can possess this structure following a major catastrophe such as a fire or large-scale clearance by man so that a large area is left to regrow.

2 Mixed age structures are typified by many animal and human age populations, where different individuals of widely varying ages are present in the same community. Unless some other factors are inhibiting regeneration, longer lived species might be expected to produce communities of this mixed age type.

The overall population may be rising, falling or in equilibrium, depending upon its present and past relationships with the varying ability of the environment to support it (the changing carrying capacity of the environment); and the nature of the interactions between the species and its competitors and associates. Populations are never completely stable through time. Leaving aside longer-term or sudden population changes due to slow climatic change or natural hazards, populations of shorter-lived animals or plants may reflect seasonal fluctuations in the environment, while longer-lived mammalian species, for example the lynx in Arctic ecosystems display amazingly regular oscillations in abundance. These cycles may be modelled and predictions made of the magnitude and frequency of future cycles.

Succession

Vegetation can itself initiate or reinforce changes as well as simply respond to changes in its environment. Here the concepts of succession and seres are important. 'Succession is defined as a directional vegetational change induced by an environmental change, or by the intrinsic properties of plants. The initial rate of vegetational change is high and subsequently falls to a low level, after which further development is governed by very slow changes of climate or physiography. This relatively stable state is referred to as the climax.' (Kershaw, 1973, p. 62).

This longer term view is most easily appreciated by considering a recently cleared landscape – the result of a landslide, dune blow-out or the excavation of a road cutting, pond or reservoir. Plants and animals in the area which are capable of quickly colonizing and reproducing will establish themselves and form pioneer communities. On land, these may stabilize the soil and change its qualities by providing organic matter which will hold water and vital nutrients such as nitrogen and phosphorus. As an example, while the fringing pioneer vegetation of a new pond may help to stabilize its banks and trap suspended sediment, it also raises the pond bed. The species thus change the properties of their habitat, a process which may continue until the environment becomes modified to such an extent that other species such as trees, which were previously unable to colonize and establish because the area was too wet or nutrient deficient, may now find the environment suitable. Once these later species have arrived, they in turn start to modify the pattern of nutrient cycling, soil conditions or the light environment. The pioneer species may then fail to reproduce and become extinct at this location.

Each successive vegetation type which behaves in this fashion is known as a seral stage. The whole process of initial vegetation development up to the final stage of relative stability or climax is known as a primary succession or prisere, as long as it is unaffected by human activity. Where man is the instigator of the original change through clearance or burning, then the vegetational changes associated with regrowth are known as a secondary sere. Good examples of priseres are relatively scarce in the developed world. Hydroseres and haloseres occur in shallow fresh or brackish water and are best observed at coastal salt marshes or at freshwater lake margins (Figure 2.8; Plate 2.1), which are gradually colonized by aquatic and semi-aquatic plants (Plates 1.1, 1.2 and 2.1). These trap sediment and contribute organic matter, building up the soil level and soil nutrient status so that eventually scrub and woodland species may invade and form a stable climax. Psammoseres occur on sands, classically on sand dune systems (although many supposed sand dune

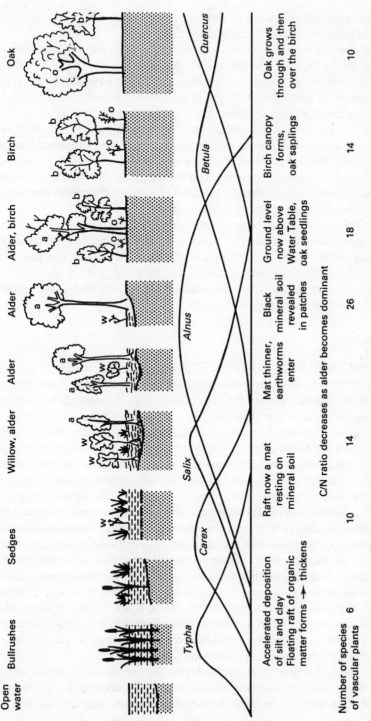

Figure 2.8 Hydrosere at Sweet Mere, near Ellesmere, Shropshire, England; note the establishment of a closed, more stable woodland cover is associated with a decrease in the number of species of vascular plants present, as those surviving in more marginal, open, wetter environments are lost (Redrawn from Cousens (1974); by kind permission of Oliver and Boyd; the original data are found in Tansley (1939); pp. 464–67.)

successions are only zonations). Lithoseres develop from bare rock, typically after landsliding or on recently exposed glacial deposits.

Marked changes have been shown to occur through the course of a succession – not only in the types of species present, but also in terms of productivity, biomass, the effect of seral stage on the environment and the overall ability of the sere to maintain its own equilibrium and resistance to change (Figure 2.8). Each stage has a greater capacity to entrap and hold nutrients for cycling within the system, with only very small amounts of nutrients being lost from mature systems compared to loss rates at earlier seral stages or from disturbed climax stages (Odum, 1983). However, the increasing ecological dominance of the many later sere colonizers, such as trees, in reducing light and habitat variety as they close the canopy space, also has the effect of reducing the overall diversity, as species preferring more open conditions are lost (Figure 2.8).

Confusions often exist between zonations and successions. Care in interpreting zonations is necessary, because the zonation may be the result of subtle and unexpected patterns of human activity and land use, both past and present, rather than successional processes. In Chapters 4, 5, 8 and 10, we make extensive reference to the study of zonations of species and communities distributed along environmental gradients.

Climax concepts

On a global scale, climate plays the all important role in determining the character of the vegetation, as shown in Figure 2.9. The stable vegetation type developing in a world vegetation region is termed the climatic climax, which emphasizes the relationship. The same type of equilibrium between soils and climate has been shown to exist over extensive areas of the world, with vegetation and soil fauna playing very important roles in linking climate and soil. These equilibrium soils are known as zonal soil types. The view of all vegetation ultimately coming into equilibrium with climate is known as the monoclimax theory. Subsequently, it has been realized that although climate still exerts a quite fundamental influence (for example, tundra or tropical rain forest will not develop in Mediterranean climates), the detailed composition of the climax vegetation and soils at the local scale will also be determined by other factors such as the topography and drainage, the properties of the soil parent material and the nature, magnitude and frequency of past and present animal and human pressures on the area. In view of this, the term polyclimax is used to emphasize the possible diversity of the vegetation types within a theoretically stable climax state for one area. The impact of human and/or edaphic factors may be substantial, for example, many of the world's tropical grasslands or savannas appear to be the result of such impact.

Where established vegetation is severely modified or partially removed, most often by man, then the resulting vegetation changes are called a secondary succession. Examples can be seen where mature woodland has been cut down.

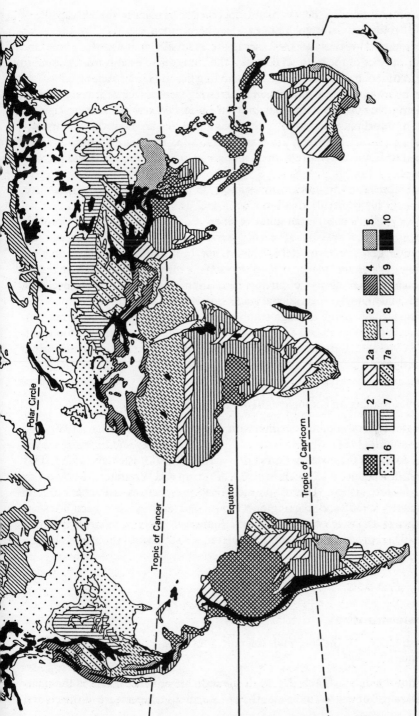

Figure 2.9 Vegetational zones (much simplified without edaphically or anthropogenically influenced vegetational regions) **A** Tropical and subtropical zones: **1** Evergreen, rain forests of the lowlands and mountainsides (cloud forests); **2** semi-evergreen and deciduous forests; **2a** dry woodlands, natural savannas or grasslands; **3** hot semi-deserts and deserts: **B** Temperate and Arctic zones; **4** Sclerophyllous woodlands with winter rain; **5** moist warm temperate woodlands; **6** deciduous forests; **7** steppes of the temperate zone; **7a** semi-deserts and deserts with cold winters; **8** boreal coniferous zone; **9** tundra; **10** mountains. (After Walter (1973); redrawn with kind permission Springer-Verlag, Heidelberg.)

The ground flora of the woodland, together with pioneer species capable of rapid colonization from refuges elsewhere, then flourish in the open conditions. However, as these areas are gradually re-invaded by tree and shrub species of the later seral stages, the climax is re-established, leading to local extinction of the more light-demanding ground flora species.

In many areas, the continuation of arresting factors such as grazing animals and cropping or harvesting may prevent succession and maintain a plagiocli-max or subclimax community. Such communities are now very common – most grasslands in Britain are plagioclimaxes maintained by grazing. If the arresting factor is removed, then succession continues towards the climatic climax.

All climax communities are subject to some change and instability, although this is usually small in magnitude. Tall trees grow old and die, fall and so produce substantial areas of open space which temporarily provide refuge for the species of earlier seral stages. In all ecosystems and communities, the term stable should not be taken as indicating that no change is taking place. All ecosystems exhibit dynamic equilibrium. Fluctuations are always occurring in relation to seasons, the life cycles of plants and animals, successional stage, nutrient cycle inputs and outputs and the passage of energy through trophic levels. All ecosystems have homeostatic mechanisms to absorb these changes and provided they are not too great, the ecosystem and communities will remain in a stable or equilibrium condition.

Further reading on ecological theory and principles

Complementary sources dealing with basic ecology are Owen (1980) and Dowdeswell (1984). More advanced reading in ecology and biogeography is to be found in the texts by Pears (1977), Krebs (1978), Ricklefs (1979), Tivy (1982), Clapham (1982), Odum (1983), Putnam and Wratten (1984) and Cox and Moore (1985). Useful physiological perspectives are contained in Bannister (1976) and Barrington (1980). Our aim has not been to write another textbook of ecological theory. Rather, it has been to stress the areas of field and practical work and their relevance to ecological problem-solving.

2.7 Projects

Classroom project

Investigation of the thermal tolerances of mistletoe (*Viscum album*) in Denmark.

1 Plot the data in Table 2.1, with the ordinate or vertical axis as the mean temperature of the warmest month, and the mean temperature of the coldest

month as the abscissa or horizontal axis. Place highest temperatures at the origin of the axis in each case.

2 Using the plotted data, draw on a line to show the climatic survival boundary conditions. Indicate why this species is better adapted to more continental climates than the holly (Figure 2.6) and using the mistletoe, demonstrate the concept of one environmental factor compensating for extreme values in another, so permitting species to survive.

Field project

The effects of plants in modifying their own environment. Assessment of micro-climatological effects of woodland, coppice and shelter belts.

Equipment

Hand-held anemometers and/or whirling psychrometer; tatter flags; light meters; environmental thermometers; drying oven; scientific balance.

Procedure

1 Relative humidity – using the transect sampling design shown in Figure 2.10, assess the impact of the woodland on wind speed and relative humidity with the aid of the hand-held anemometers and whirling psychrometers.

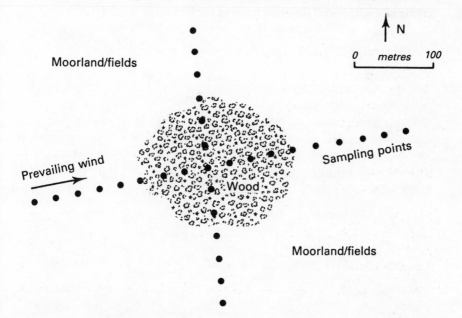

Figure 2.10 Suggested layout of survey points to investigate the microclimatological effects of woodland, coppice and shelterbelts

Transects should be laid out well beyond the limits of the small woodland, coppice hedgerow or shelter belt(s) chosen for study.

2   Temperature – carry out a similar experiment for air temperature.

3   Exposure – this may be assessed by using tatter flags which were developed by Lines and Howell (1963), who observed that flags in exposed locations in Scotland became frayed and hence lost weight when compared to flags from more sheltered situations. They used this idea to identify less exposed locations for forestry plantations. The cloth used by Lines and Howell is not readily available, but other material may be substituted. Flags may be made by cutting pennants 0·4 m long of standard size and weight. These should be cut in an identical manner in relation to the thread of the cloth. Old linen sheets of the same material are suitable. The flags may then be attached to 1 m high wooden or bamboo stakes, having first been weighed dry. Flags are then left out along the transect for 1–3 months during the least favourable times of the year. Relative exposure can then be calculated by taking the dry weight of the flags after exposure and calculating the weight loss as a percentage.

4   Light intensity and spectral quality is changed greatly by tree crown foliage. Use the light meter to assess variations in light intensity in transects across the woodland/coppice/shelter belt. Spectral quality may also be investigated by repeating measurements with filters over the light meter, either singly or in combination.

5   All these surveys may be carried out at a range of different locations, at varying altitudes and at different times of the year.

3 Making a start

3.1 Formulating ideas

Many students experience considerable difficulty in formulating ideas for practical and project work in ecology, biogeography and environmental biology, yet problems and potential projects are to be seen everywhere. If such difficulties are found, then there is no substitute for a visit to a site or several sites of possible interest, such as a nearby area of semi-natural vegetation, nature reserve or beauty spot. Within town and city boundaries a surprising number of ideas may be generated. City parks, derelict and abandoned land, back gardens, old railway lines, even walls and pavements all have considerable ecological interest.

Once a site has been identified, a series of questions may be asked:

1 what plants and animals are to be found there?
2 where are they and how are they distributed?
3 what are the possible factors limiting their distribution?
4 can plant communities be recognized with particular associations or groups of species?
5 are there other similar sites nearby? How do they compare?
6 what is the history of the area? What changes have occurred in the recent past?
7 what is the successional status of the site?
8 is man exerting an important influence on the plants and animals?
9 is the site worth conserving and keeping as it is or are there possibilities for change? Are the plants and animals under threat from man's activities?
10 what is already known about the site and all that it contains?

As these questions are posed, a series of problems which require investigation should suggest themselves.

A second approach is to locate an environmental gradient, such as change in altitude up a hill slope, a land–water margin by a lake or by the coast, or a gradient of man's impact such as the effects of trampling around a heavily used recreation area. Again the same questions may be posed and projects generated.

Having decided on a problem, try to formulate two types of survey.

1   Those which are necessary simply to describe the nature of the site – such as listing, assessing the abundance and mapping the distribution of the plants and animals
2   Those which result from the formulation of hypotheses (Section 1.3).

Next, try to find out what information is already available about either the site, the plants and animals found there, or the ecological relationships around which hypotheses may have been generated. If some research has already been published, examine the work and ask the following questions.

1   Do the ideas and explanations agree with your own observations?
2   Are there important aspects of the study which have been neglected?
3   Can you improve on the study?
4   What techniques and sampling methods have been employed? Can these be improved or adapted?
5   Can the study be profitably repeated, perhaps in a different location or at a different time of year?
6   Is it possible to take the ideas and approaches used in the published study and apply them, perhaps in modified form, to a different problem in a different environment?

Many good topics, which can be adapted or repeated in different situations are to be found in the following journals: *Journal of Biological Education*; *Geography*; *Teaching Geography*; *Field Studies* and *School Science Review*.

The more advanced scientific journals are also worth reading in this context: *Applied Geography*; *Biological Conservation*; *Ecology*; *Ecological Monographs*; *Environmental Conservation*; *Journal of Biogeography*; *Journal of Applied Ecology*; *Journal of Ecology*; *Journal of Environmental Management*; *Environmental Pollution*; *Urban Ecology*.

3.2   Identification of species

The problems of species identification are probably the biggest deterrent to students undertaking ecological project work. In practice many of these quite natural anxieties are unnecessary. First, interesting and worthwhile projects can be carried out on subjects such as heather, bracken, common trees and well-known species of birds and rodents. Second, many relatively inexpensive pictorial guides to most groups of plants and animals specifically designed for non-specialists are now available (see Appendix).

Identification of specimens in the field is not always possible. Such species can always be photographed and the photograph taken to a local museum or expert. In the field, each species can be given a name and as long as that name is used consistently until the correct one is found no error should result.

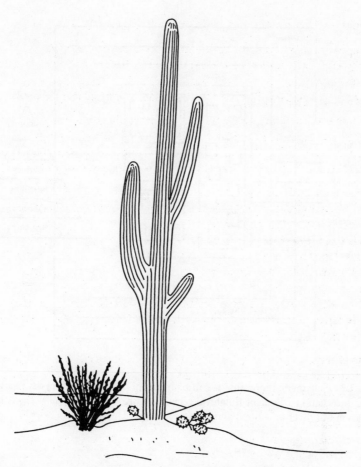

Figure 3.1   A sketch of *Cereus giganteus* in the Sonoran Desert, USA

3.3   The Latin binomial and naming of species

Fans of western films shot in the deserts of southern Arizona will recognize the Saguaro cactus, shown in Figure 3.1. The saguaro lives in the Sonoran Desert of Arizona and Mexico (Steenbergh and Lowe, 1977). Its Latin name or the Latin binomial is *Cereus giganteus*. This scientific name is constructed like the name of a person, except that the surname is first. The saguaro is of the genus *Cereus* and the species *giganteus*. Four other species of *Cereus* occur in Arizona – *Cereus thuberi*, *Cereus schotti*, *Cereus greggii* and *Cereus striatus*. In everyday speech, the common name of saguaro is applied to all of these five species and as a result causes unnecessary confusion. The use of the binomial is very important in avoiding such misunderstanding. In the British Isles, the Scots heather or Ling (*Calluna vulgaris*) is a separate genus to the true heathers of the

Categories

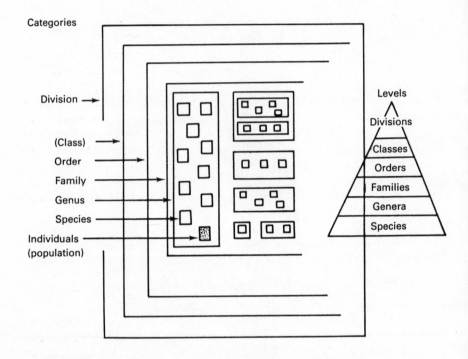

Figure 3.2   Schematic representation of plant categories (taxa) and levels, recognized by The International Code of Botanical Nomenclature. (After Tivy (1982); with kind permission of Oliver and Boyd and Professor V. H. Heywood.)

genus *Erica*. The Scots bluebell (*Campanula rotundifolia*) is not the same as the English bluebell (*Endymion non-scriptus*). *C rotundifolia* is called a harebell in England.

The hierarchy of taxonomic nomenclature is shown in Figure 3.2 and classification of a species is completed by taxonomic specialists who follow a distinct set of rules for naming a species – the International Code for Botanical (or Zoological) Nomenclature. In principle the code is very simple.

1   One taxon – a single taxonomic unit of any scale (e.g. a variety, a species, a genus) may only have one valid name;

2   More than one taxon may not have the same name;

3   In cases not in agreement with the first two points, the decision of which name is to be used is based upon priority of publication;

4   The identity of the plant/animal to which a name has been given is according to the type specimen designated by the author as representative.

In practice, such matters often become complicated subjects for taxonomists and thus of limited interest to students of ecology and biogeography, who simply have to try to remember the names. Unfortunately, however, names do change from time to time. This has just occurred for a number of grass species in Britain of which the most widely known is the common bent, which was formerly known as *Agrostis tenuis* until 1981, when in the latest Excursion Flora of Clapham, Tutin and Warburg, it was changed to *Agrostis capillaris*. The exact reasons for this probably lie in 3 above with the name *capillaris* having been given to the species elsewhere at an earlier date, but the relationship between the two having only just been realised.

A further indication of the difficulties involved in taxonomy is given by the quotation from Benson, *The Cacti of Arizona* (University of Arizona Press 1969), p. 108, when referring to *Cereus*:

'Cereus undoubtedly needs to be subdivided but this will become feasible only after laborious and very extensive gathering of data. The presently available information is inadequate and the way forward is not clear. Engelmann noted he may be able to classify Cereus in the next world. The writer may be in the same position.'

Previous scientists regarded the saguaro cactus as sufficiently different to the other members of the genus *Cereus* to merit the cactus being placed in a new different genus – *Carnegia* – in honour of the Scottish steel magnate Andrew Carnegie (from whom they hoped to obtain a research award). Others have also changed the species name. Consequently, when consulting the standard text or Flora for the area, or the group of plants, it is necessary to inspect the synonomy, which sets out all the various names or synonyms for a particular species. For example:

Synonomy
Cereus giganteus Engelmann.
Pilocereus engelmannii Lemaire. nom nov.
Pilocereus giganteus Rumpler.
Carnegia gigantea Britton and Rose.
Type locality – Coolidge Dam, Arizona.

The list sets out the various names given by botanists studying this cactus. The tradition of naming species after the scientists who first found them has now largely disappeared; today new names reflect the characteristics of the species or its habitat.

3.4 Basic sources of information

Floras and Faunas

The Flora or Fauna of an area are the first basic sources which merit attention in any project. These may take the form of lists of the plants or animals found in the locality, or the plant and animal communities present, often with further notes on how to identify the various species and on their most characteristic habitats. While the plants and animals constitute the flora and fauna of an area, a published Flora and Fauna should be a catalogue which has been systematically compiled using the hierarchical structure shown in Figure 3.2 and where species can be identified by use of a dichotomous key based on certain physiognomic properties. The present *Flora for the British Isles* is Clapham, Tutin and Warburg (1962; 1981) and the Flora for Europe, *Flora Europaea* is still being published in a series of volumes.

The first book with a claim to be a County Flora for the British Isles was published as long ago as 1660 by the naturalist John Ray. It dealt with the county of Cambridgeshire and was reproduced in facsimile form by Ewen and Prime (1975).

Case Study: The Flora of Essex

The first Flora of Essex was published in 1862 by G. S. Gibson, incorporating over three centuries of observations made by earlier botanists and herbalists. One of Gibson's principal sources was the *Herbal of Gerard*, published in 1597. Gerard's Herbal was written primarily as a guide for the medical profession and sets out the medicinal and other properties of plants. Gibson combined this and other data sources and set out lists of species and habitat notes, having divided Essex into eight arbitrary sub-regions for descriptive purposes.

The twentieth-century *Flora of Essex* by Jermyn (1974) is much more sophisticated. The county has been divided into fifty-seven 10 × 10 km grid cells and the presence, absence or local extinction of 1733 species, sub-species, varieties and hybrids are presented on this grid for the whole county. For comparative purposes, detailed maps of preferred habitats, geology, climate and land use are presented. Study of the two Floras indicates that 30 species have become extinct in the county during the period 1597 to 1862; 7 species became extinct in the period 1862 to 1930 and a further 23 species have been eliminated between 1930 and 1974.

Photographic, map and archive sources

A wealth of background information for any project may be obtained from photographic, map and archive sources. Museums, libraries and public records offices contain information from old maps and photographs, estate records and early accounts of the flora and fauna. School, college and university libraries and departments will also contain more recent information as maps, books, journals and computer data banks. Permission must usually be sought to visit many such institutions but most are willing to help. Evidence from such historical and contemporary sources can be used to examine the changing nature of the vegetation and animals of an area over time, and to determine the extent to which modification of ecosystems may have occurred due to man's activities.

An interesting example of the use of historical records in Australia is the *Map of the Forest Trees of Victoria*, published by the Victorian Department of Mines on October 24, 1866, which was reprinted at half-size reduction with valuable notes in the *1978 Victorian Yearbook* (Knight, 1978). The map was compiled from data obtained by Baron Von Mueller who traversed the state in his capacity as government botanist. During his visits he had the habit of scattering blackberry seeds which he hoped would provide food for lost peasants and birds, little realizing the ecological havoc the blackberry would later cause in many parts of the Victorian bushland. Land surveyors in these colonial regions frequently recorded the vegetation in detail because the plants of the area and their distribution provided valuable information on the likely productivity of the land. An example of the early work of these surveyors is shown in Figure 10.3(b). This can be compared with the results of satellite mapping (Chapter 10).

A bibliography of vegetation maps for different parts of the world has been produced by Professor A. W. Küchler (1965, 1966, 1969, 1970). Sheail (1980) has also produced an excellent brief guide to historical sources for ecologists and biogeographers in Britain.

Maps of flora and fauna

Compared to the large number of vegetation maps, few maps of individual species distributions of floras and faunas have been produced. In Britain, the best examples are the *Atlas of the British Flora* by Perring and Walters (1962) and the *Atlases of the British Fauna* e.g. Kerney (1976) which are published by the Institute of Terrestrial Ecology (see Appendix).

Interpretation of these maps is greatly aided by the comparative environmental data which are available as overlays from the ITE. An example of a typical map is shown in Figure 9.4.

3.5 Projects

Classroom project

An investigation of the variation of higher plant species numbers in each 2 × 2 km grid cell or tetrad of the *Flora of Rutland* (Messenger, 1971).

The county of Rutland was the smallest of the former English counties and was incorporated into the county of Leicestershire during the local government reform of 1973. The area is mainly rural with moderate relief. Maps of the local geology, topography and drainage and the abundance of species in each grid cell (tetrad) are presented in Figure 3.3.

Equipment
Figure 3.3; tracing paper; transparent acetate overlay; graph paper; drawing pens; Ordnance Survey maps 129, 130, 141 at 1:50,000 or 122, 123, 133 at 1:63,360.

Procedure
1   Study the following hypotheses:
(a) the number of plant species in each tetrad is a direct function of the area of land surveyed in each cell (see Figure 3.3);
(b) the flora is significantly richer per unit area in the east of the county and poorest in the north-west, reflecting variations in geology;
(c) variations in species abundance largely reflects the variety of soil types and the frequency of aquatic habitats within each cell.
2   Hypothesis (a) – using data on species numbers and the area surveyed in each tetrad of Figure 3.3, plot a graph of species numbers against area surveyed in each cell.
Hypothesis (b) – transfer species data to a transparent overlay. Investigate the first part of the hypothesis by plotting a graph of species numbers per tetrad against its distance east of a point A along the west-east baseline A-B. Investigate the second part of the hypothesis by using a transparent overlay and calculating the average number of species within each tetrad which has over 50 per cent of its area covered by (i) Jurassic limestone; (ii) the Jurassic clays; (iii) marlstones; (iv) Northampton sands.
Plot the results as a series of bar graphs (as in Figure 6.5).
Hypothesis (c) – prepare an overlay showing the number of different soil types plus the number of aquatic habitats in each tetrad. A standard method for identifying the number of aquatic habitats will have to be devised using the information on drainage and any other information gained from the 1:50,000 Ordnance Survey map. Investigate the hypothesis by plotting a graph of the number of species per tetrad against the number of soil types and aquatic habitats per tetrad.
Comment on the extent to which your results support or refute the hypotheses.
3   Using the maps for *Clematis vitalba* and *Ophioglossum vulgatum* and the 1:50,000 Ordnance Survey map, examine the relationships between man's activities in the form of urban development and agriculture, and the distributions of the two species.

Field projects

Literature and baseline data search
As preparation for later projects suggested in this text, visit your local

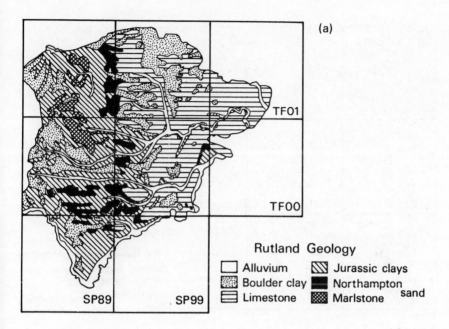

(a)

TF01

TF00

SP89   SP99

Rutland Geology

☐ Alluvium   ▨ Jurassic clays
▦ Boulder clay   ■ Northampton
▤ Limestone   ▩ Marlstone   sand

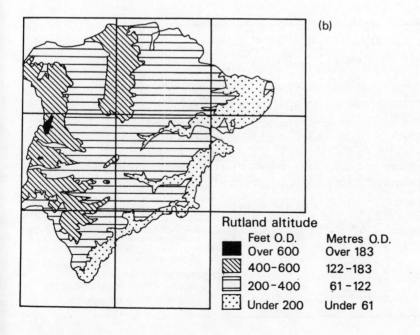

(b)

Rutland altitude

	Feet O.D.	Metres O.D.
■	Over 600	Over 183
▨	400–600	122–183
▤	200–400	61–122
⦂	Under 200	Under 61

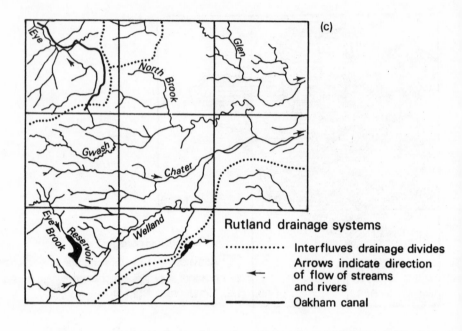

(c)

Rutland drainage systems

················ Interfluves drainage divides

←· Arrows indicate direction of flow of streams and rivers

───── Oakham canal

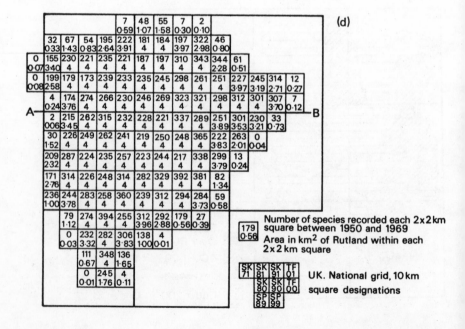

(d)

Number of species recorded each 2×2km square between 1950 and 1969

Area in km² of Rutland within each 2×2 km square

UK. National grid, 10 km square designations

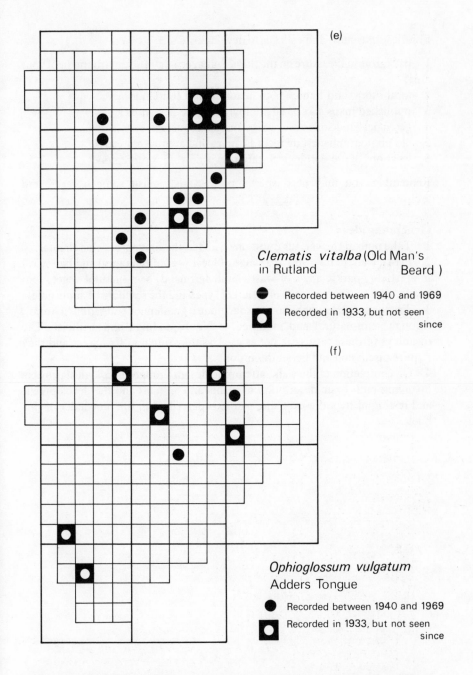

Figure 3.3 Maps from the flora of Rutland: (a) geology; (b) relief; (c) drainage; (d) plant species numbers in 2 km × 2 km grid squares; (e) species distribution of *Clematis vitalba* (Old Man's beard); (f) species distribution of *Ophioglossum vulgatum* (Adder's tongue) (Redrawn with kind permission from Messenger (1971).)

libraries, museums and colleges to develop lists of:

1   background literature on the flora, fauna, ecology and environment of your area;
2   local Floras and Fauna, both historical and contemporary;
3   published maps and aerial photographs (see Appendix);
4   geological and soil survey maps and data sources;
5   old photographs, estate records, maps and archive materials;
6   locations of meteorological stations within your local region.

Remember you may need special permission to visit some libraries and archives.

Generating ideas

1   Take note of the various questions listed in the first part of this chapter.
2   Plan a walk to take you to a range of local sites of ecological and biological interest e.g. parks, areas of waste land, farmland, semi-natural vegetation, beauty spots, streams, roads of different types and the seashore (if applicable).
3   Try to place these habitats in order along a gradient or scale from 'natural', through 'semi-natural' and 'modified' to 'highly modified' and 'artificial'. The meanings of these terms are not as clear as they might at first seem and they require discussion and better definition.
4   On completion of the walk, attempt to design a range of ecological projects involving both basic description of plants and animals and the development and testing of hypotheses, using methods described in the remainder of this book.

Part Two

Field Surveys of Plants and Animals

4 Describing vegetation in the field – physiognomic methods

4.1 Introduction

The survey and description of plant and animal species and communities in the field form an essential part of many biological and environmental projects, although for various reasons, many people have tended to concentrate on plants. Vegetation is often the most conspicuous feature of terrestrial ecosystems and provides the habitat within which animals live. Although seasonal variations in yield, performance and biomass occur (see Chapter 6), in the very short term, vegetation may be seen as static and hence is far more easily measured than animals, which are mobile and frequently elusive. A large variety of methods are available for the description of both plant and animal resources in the field. However, whereas the plants in one area may usually all be described using the same or very similar techniques, with animals, the methods applied are more varied, depending on the individual species or group under study.

4.2 Describing vegetation

Selection of method

The choice of method for vegetation description depends on a number of important factors.

1   The aim of the survey – depending on the purpose of a particular project, different attributes of plants and vegetation will need to be described.
2   The scale of study – description methods vary greatly, according to whether large areas of, for example, thousands of square kilometres are being described, or whether very detailed descriptions are required in a restricted area of a few square metres.
3   The habitat – some techniques are only suited to a particular habitat or vegetation type. As an extreme example, the methods required in a chalk

grassland in southern England may be very different from those in a tropical rain forest in Northern Australia or South America.

Two other questions also require careful consideration; whether or not it is necessary to identify the plant species; and what methods of sampling should be used? Methods for describing the physiognomy and structure of the vegetation are described first in this chapter as they do not, in general, require the identification of all the species within an area nor a rigorous sampling design. This is followed in Chapter 5 by floristic methods for vegetation description, where accurate identification of all species present is essential, and problems of sampling must be considered.

4.3 Physiognomic methods

Approximately 300,000 species of flowering plants have been identified in the world. As a result, even expert botanists, when surveying parts of the world with which they are unfamiliar, experience major problems with identification and description.

Appropriate and accurate reference books may be unavailable, or in an unfamiliar language. Surveys of remote and often large areas will also invariably involve field data collection by surveyors who cannot be expected to know many of the species present. At this scale, much mapping will be carried out using aerial photographs or satellite imagery on which individual species will almost certainly be unidentifiable. The data required are often collected for economic or conservation purposes and while species identification is undoubtedly helpful, the size, shape and density of the vegetation will often be more important. Such attributes are readily measured by non-specialist surveyors and students.

To meet these requirements, physiognomic methods of vegetation description are often used. These are based on the size, shape, and form of the vegetation and do not demand a detailed knowledge of species composition. As a result, they are relatively easy to apply and form an excellent introduction to vegetation survey for biology and geography students.

In this section, physiognomic survey methods appropriate to small areas are described first, followed by techniques capable of dealing with the vegetation of very large areas up to the size of continents.

4.4 Describing stratification

Profile charts

Most people will have noticed that in local woodlands, parks, or in the countryside in general, the vegetation often comprises a series of horizontal layers involving particular groups of species. This layering is referred to as stratification, and may be described by a profile chart showing the type, height and position of each plant species along a survey line. Tropical rain forest is characterized by its complex stratification, with five or more tree and shrub layers which are shown by a profile chart in Figure 4.1.

Such forest is commonly divided into five layers:

A   the emergents, the tallest trees present in the area;
B and C   two strata of canopy tree species;
D   the shrub layer;
E   the lowest herb/sapling stratum.

This pattern of organization enables trees and shrubs to maximize their exploitation of the incoming light in this productive and competitive environment.

Stratification is one of the most characteristic features of vegetation. On a world scale it is important for the differentiation of world formation types such as the multi-layered tropical rain forest, and tundra which may have just one or two layers. Stratification is also important at smaller scales. Temperate oak woodland is often said to comprise four layers. The uppermost tree or canopy layer lies above an understory, comprising shrub and tall saplings which form the shrub layer. The field or herb layer of non-tree herbaceous plants may reach up to 2 m in some cases, and in turn covers the ground or moss layer (see Section 7.4). The stratification continues into the soil with roots of species preferentially occurring at different levels.

The profile chart may be used to study the manner in which vegetation stratification alters as plants have to cope with change along an environmental gradient, such as the margins of a lake, river or track. The short profile charts in Figure 4.2 have been extended into three-dimensional pictures of river bank vegetation. When floristic data describing the actual species present are also collected, the profile chart effectively becomes a transect diagram (see Section 5.6).

Although stratification is a very important property of vegetation, layers may on occasion grade into one another, especially in some tropical forests. Nevertheless, in most situations some layering in vegetation is readily apparent.

Methods for small area surveys
Equipment: 30 m tape; string; marker pegs; optional clinometer

Procedure
1   Lay out a rectangular strip or single line through the chosen sector of vegetation. If in forest, ideally 8+ m wide and at least 50 m in length.
2   Construct a profile chart by accurately sketching the plants to scale along the profile line.

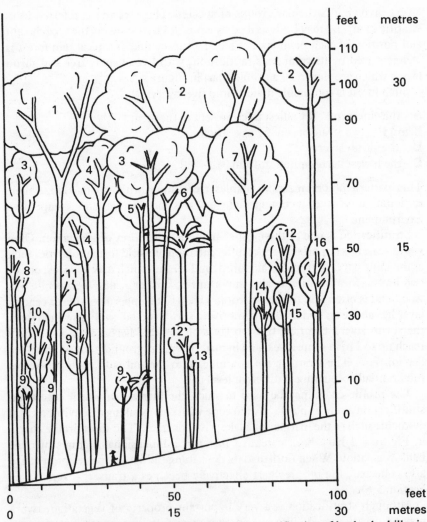

Figure 4.1 Profile diagram through the complex stratification of lowland – hill rain forest; Ramu-Madang, Papua New Guinea

Tree species:

1	*Tristiropsis*	9	*Litsea*
2	*Intsia bijuga*	10	*Syzygium*
3	*Pimeleodendrom amboinicum*	11	*Homalium*
4	*Canarium*	12	*Diospyros*
5	*Pometia pinnata*	13	*Ficus*
6	*Maniltoa*	14	*Gonocaryum*
7	*Laportea*	15	*Microcus*
8	*Myristica*	16	Annonaceae

Shrubs and ground plants are omitted. (Modified and redrawn from Robbins *et al.* (1976); by kind permission CSIRO Australia.)

(a)

Typha latifolia

Sparganium eurycarpum

Scirpus validus
Zizania aquatica
Nuphar variegatum
Saggittaria cuneata
Lemna minor
Potamogeton nodosus
Heteranthera dubia
Ceratophyllum demersum
Elodea nuttallii

(b)

Phragmites communis

Hibiscus moscheutos

Pontederia cordata
Peltandra virginica
Saggitaria latifolia
Nuphar advena

Polygonium hydropiperoide

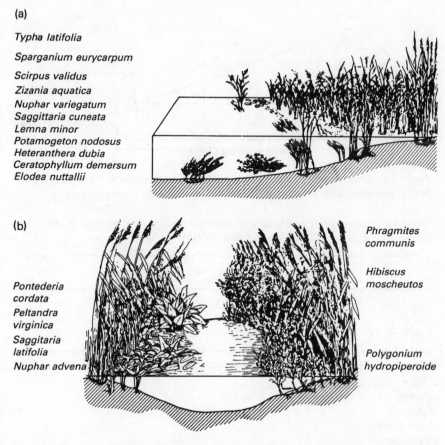

Figure 4.2 Profile charts extended into three dimensional diagrams showing the stratification of vegetation: (a) in a species-rich, slow flowing stream with a wide swamp fringe, (Mukwonago River, Wisconsin); (b) in a river with *Phragmites communis* marsh behind clumps of wide-leaved emergents (Delaware, USA) (Redrawn from Haslam (1978); with kind permission of Cambridge University Press.)

3 With trees, measure tree/shrub trunk diameters, height to first large branch, lower limit of crown, crown width, and total height.

These opportunities arise during forest clearance or thinning. A simple method of determining heights of trees or lowest branches is shown in Figure 4.3. It is possible to purchase clinometers which automatically indicate the tree height when 30 m from the tree base.

Large area surveys using profile data
For large area surveys, data on the abundance and spatial distribution of each species may be obtained with information on stratification using the methods

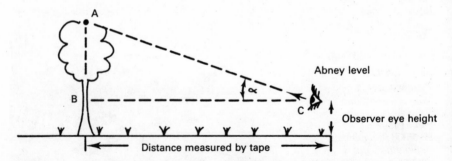

Figure 4.3   Determination of tree height with a clinometer held at eye level. Distance B–C is measured with a tape: tree height = B–C tan α + observer height

Table 4.1   *Scheme for recording data on vegetation stratification, abundance and spatial distribution developed by Christian and Perry (1953)*

	Trees	Shrubs	Herbs
Tall	A_3	B_3	C_3
Medium	A_2	B_2	C_2
Low	A_1	B_1	C_1

Density	Suffix	Heights record mean value for each stratum
Dense	x	
Average	y	
Sparse	z	
Very dense	xx	
Very sparse	zz	

The data is written $A_3^{50}y$, indicates a tall tree stratum, 50 m high on average and of average density.

developed by Christian and Perry (1953) set out in Table 4.1. Their technique is a very rapid shorthand notation system. In Figure 4.1, the highest stratum of trees would be defined as $A_3^{30}y$.

4.5   The Dansereau structural description method

In addition to observations of plant size and ground coverage, Dansereau (1957) incorporated information on life-form, plant function and leaf size, shape and texture into structural vegetation description. The term life-form refers to the growth form of the plant and often shows a distinct relationship with environmental factors. For example, many high altitude trees exhibit a creeping growth habit called 'kromholz' when they are near the limits of their

Table 4.2   *Six categories of criteria to be applied to Dansereau's structural description of vegetation types* (Adapted and redrawn from Dansereau (1957).)

1. LIFE FORM			4. FUNCTION		
T	◯	Trees	d	☐	Deciduous
F	♀	Shrubs	s	▥	Semideciduous
H	▽	Herbs	e	▦	Evergreen
M	⌢	Bryoids	j	▩	Evergreen – succulent;
E	☆	Epiphytes			or evergreen – leafless
L	🖎	Lianas			

2. SIZE			5. LEAF SHAPE AND SIZE		
t	Tall	(T: minimum 25m)	n	⌖	Needle or spine
		(F: 2–8m)	g	◊	Graminoid
		(H: minimum 2m)	a	◇	Medium or small
m	Medium	(T: 10–25m)	h	△	Broad
		(F, H: 0·5–2m)	v	♡	Compound
		(M: mimimum 10cm)	q	⊙	Thalloid
l	Low	(T: 8–10m)			
		(F, H: maximum 50cm)			
		(M: Maximum 10cm)			

3. COVERAGE		6. LEAF TEXTURE		
b	barren or very sparse	f	▨	Filmy
i	discontinuous	z	⊞	Membraneous
p	in tufts or groups	x	■	Sclerophyll
c	continuous	k	▦	Succulent; or fungoid

tolerance, owing to the severely exposed environment. However, in less stressed circumstances below the tree line, they may have a normal tree form.

Quite unrelated species from unrelated families may also have evolved the same life-form. For example stem-succulents of similar growth form have emerged in the unrelated families of the *Cactaceae* (the Cactus family) and the *Liliaceae* (the Lily family). Similarly, particular types of vegetation are characterized by either deciduous, evergreen or evergreen succulent leaves and again this represents a response to environmental conditions.

In 1957 Dansereau published the notation scheme shown in Table 4.2. This information is included in a symbolized form on profile diagrams. No actual information on species identification is required, allowing the technique to be used rapidly by inexperienced surveyors working over large areas. An example of its use in East African grasslands and savanna is shown in Figure 4.4, and for a British woodland in Figure 13.4. Such data are of particular value in remote areas where, for example, road or communications links are being developed.

4.6   Describing world vegetation formations

Description of the vegetation of the world or major continents necessitates the use of terms and methods which will have meaning to the scientist in South America, as much as in Europe, Asia, or Australasia. Structural formations are

Grassland without trees

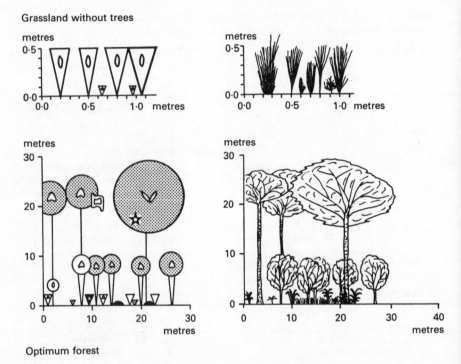

Optimum forest

Figure 4.4   Use of Dansereau's method of vegetation description to portray grassland and woodland in Zimbabwe (Modified from Tomlinson (1973).)

used for the primary sub-division of the world's vegetation. It has often been noted that where similar environmental conditions occur in widely separated parts of the world, the structural formations found are frequently similar. For the scientific purposes of describing and mapping the vegetation of continents and the world, and identifying those major factors which seem to be controlling the distribution of its various components, a number of structural classifications of world vegetation have been developed.

The Fosberg–IBP scheme

The International Biological Programme (IBP) adopted Fosberg's second approximation to a structural classification of world vegetation (Peterken, 1967). The method is based on strictly vegetational criteria, deliberately avoiding incorporation of environmental data. It is summarized in Table 4.3 which reveals a primary subdivision based upon the spacing or cover of the major components of the plant biomass.

1   Closed vegetation – crowns or shoots interlocking
2   Open vegetation – crowns or shoots not touching

Table 4.3 *Summary of Fosberg's formation system* (After Peterken (1967).)

Legend:
- ■ Closed
- O Open
- S Sparse
- x Absent to closed
- ao Absent to open
- s Absent to sparse
- (blank) Absent

	Floating aquatic	Submerged aquatic	Bryoid	Broad leaved herbs	Short grass	Tall grass	Dwarf shrub	Shrub	Tree
CLOSED VEGETATION									
1 A Forest			x	x	x	x	x	x	■
B Scrub			x	x	x	x	x	■	
C Dwarf scrub			x	x	x		■		
D Open forest with closed lower layers			←——— Closed ———→						O
E Closed scrub with scattered trees			x	x	x	x	x	■	S
F Dwarf scrub with scattered trees			x	x	x		■		S
G Open scrub with closed ground cover			←—— Closed ——→					O	
H Open dwarf scrub with closed ground cover			←— Closed —→				O		
I Tall savanna			←— Closed —→				s	s	S
J Low savanna			←Closed→			s	s	s	S
K Shrub Savanna			←—— Closed ——→				←—S—→		
L Tall grass			x	x	x	■			
M Short grass			x	x	■				
N Broad leaved herb vegetation			x	■	ao	ao			
O Closed bryoid vegetation			■	s	s				
P Submerged meadows	ao	■							
Q Floating meadows	■	x							
OPEN VEGETATION									
2 A Steppe forest			ao	ao	ao	ao	ao	ao	O
B Steppe scrub			ao	ao	ao	ao	ao	O	
C Dwarf steppe scrub			ao	ao	ao		O		
D Steppe savanna			ao	ao	←—O—→		ao	ao	S
E Shrub steppe savanna			ao	ao	ao	ao	s	S	
F Dwarf shrub steppe savanna			ao	ao	O		S		
G Steppe			←——— O ———→						
H Bryoid steppe			O						
I Open submerged meadows	s	O							
J Open floating meadows	O	s							
SPARSE VEGETATION									
3 A Desert forest			s	s	s	s	s	s	S
B Desert scrub			s	s	s	s	←—S—→		
C Desert herb vegetation			←——— S ———→						
D Sparse submerged meadows		S							

3 Sparse vegetation – crowns or shoots separated on average by more than the plant's crown or shoot diameter.

The second subdivision is into units termed formation classes. Here primary consideration is given to differences in the heights of the vegetation layers and their continuity or discontinuity. At least one of the layers in the vegetation must be continuous or closed to distinguish members of the closed formation class, from those of the open structural grouping.

The third division is on the basis of dominant life form, with an emphasis on leaf texture, leaf shape and thorniness. Peterken gives further definitions of all technical terms used in this scheme.

The Australian structural formation scheme

Some plant ecologists were unhappy with the Fosberg–IBP scheme, and co-operated to produce a descriptive scheme for mapping the vegetation of the entire continent of Australia that was both easier to understand and simpler to use in the field (Specht *et al.*, 1974). The excellence of their surveys are shown in the published maps – *The Natural Vegetation of Australia* (Carnahan, 1976) and the *Vegetation Map of Papua New Guinea* (Paijmans, 1975). Both maps with accompanying explanatory reports are readily obtainable (see Appendix). A section of the first map dealing with much of South Australia is redrawn here as Figure 4.5(a). A recent map of mean annual precipitation throughout this area is shown in Figure 4.5(b). Precipitation is a major limiting factor for vegetation and economic development in this state – the driest state of the driest continent.

This Australian scheme is very easy to use. It emphasizes the photosynthetic and ecological dominance of the principal vegetation layers present but expands the basic physiognomic classification to include readily obtained floristic data. The data are fairly easily collected in the field by non-specialists. A major asset lies in the extent to which the method can be used to describe study areas of all sizes, up to those of continental dimensions and can be readily integrated into basic resource-gathering survey programmes which invariably involve extensive use of aerial photography, supported by limited ground survey (Chapter 10).

The scheme is set out in Table 4.4. The initial division is based on the height and life-form of the dominant plants into trees, shrubs, or herbs; the second division on the percentage foliage cover of the ground by the dominant layer in the vegetation; the third division is based on recognition of the genus or family that typifies or characterizes the vegetation of the principal vegetation layer. The method can be illustrated by an example.

Eucalyptus tall shrubland with low shrubs (eS2Z) is widespread in South Australia (Figure 4.5(a)); a mapped distribution which may be compared with Landsat satellite imagery of the Adelaide area shown on the back cover The code eS2Z is interpreted as follows:

Tallest stratum		Lowest stratum		
genus	growth form	foliage cover	genus	growth form
e = *Eucalyptus*	s = tall shrub 2 m high	2 = between 10 and 30 per cent	undifferen- tiated	low shrubs

The tallest stratum is usually dominated by mallee forms of *Eucalyptus* or sometimes by *Acacia* species, although in some cases there may be a range of dominant genera. This vegetation type is widely distributed in South Australia, mainly where annual precipitation is in the range 200–450 mm; and especially on solodized brown soils and solodized solonetz soils (Carnahan, 1976; Bridges, 1978).

4.7 Projects

Classroom projects

Stratification of vegetation
1 Describe the structure of the plant community with the following notation in the scheme of Christian and Perry:
$A_3^{55}z$; $A_2^{40}y$; $A_1^{15}x$; $B_3^{9}z$; $B_2^{4}z$; $B_1^{1}y$; $C_2^{1}zz$; $C_1^{0.5}zz$.
2 Using the same notation describe the profiles for a tropical rain forest shown in Figure 4.1. The vegetation in all layers is of average density.
3 Tropical rain forests are often described as being composed of three main tree layers (A, B, C) above the shrub layer (D) and lowest herb stratum (E). Investigate this proposition by constructing a histogram showing the number of intercepts of tree and shrub crowns from a series of horizontally drawn sampling lines at 2.5 m vertical intervals up through the profiles diagram of Figure 4.1, illustrating the lowland rainforest of Papua New Guinea.
Do three distinct peaks – modal values – appear on the histogram? Are there any differences between the numbers of species represented at each layer of the stratification?
4 Compare the results from 3 with those obtained by repeating the exercise on the coppiced elm woodland of Hayley Wood, Cambridgeshire (Figure 4.6). Compare and contrast your results for each profile and list possible causes for the observed differences.
Coppicing is a means of exploiting woodlands containing hardwoods, whereby an open network of larger trees is maintained to provide plenty of space for shrubs such as hazel (*Corylus avellana*) and other underwood, grown for periods of 10–15 years, before being cut down to provide timber for stakes and hurdles.

Field projects

1 Investigate how and why stratification alters along a natural environmental

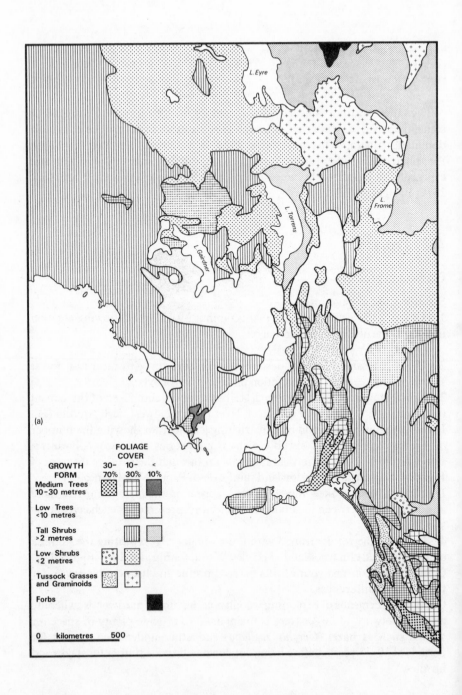

Figure 4.5(a)   Map of the vegetation of South Australia (Redrawn from Carnahan (1976); with his kind permission and the Division of National Mapping, Canberra, Australia.)

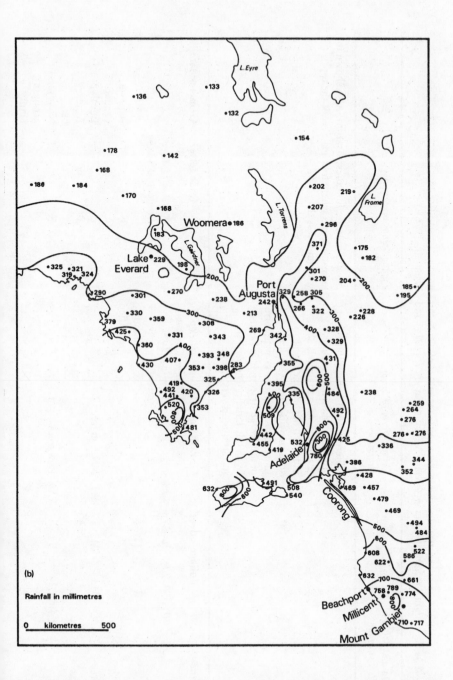

Figure 4.5(b) The distribution of precipitation in South Australia (Reproduced from unpublished data, by permission of the Bureau of Meteorology, Adelaide, South Australia.)

Table 4.4   *The Australian structural formation scheme (After Specht et al. (1974) and Carnahan (1976).)*

Growth form and foliage cover of tallest stratum	Growth form of lower stratum	Typical genus or family
	Where growth form of the lowest stratum is shown as trees or shrubs the growth form is also indicated of the next tallest stratum with foliage cover of more than 10%. The growth form of this lower stratum is indicated by a code letter only unless the top stratum is very sparse.	Always indicate for tallest stratum. Also shown for lower stratum where foliage cover of the top stratum is less than 10%

Growth form	Foliage cover				Growth form		Code letter		Genus or family
	70%	30–70%	10–30%	10%	*If Top Stratum 10% cover*	*If Top Stratum 10% cover (slightly modified)*	*Top Stratum*	*Lower Stratum*	
							b		Banksia
							c		Casuarina
						Trees and Shrubs	e		Eucalyptus
							W	w	Acacia
									etc.
							X	x	No typical genus or family
Tall Trees 30 m		T3	T2						
Medium Tree 10–30 m	M4	M3	M2	M1	Medium tree	M			
Low Trees 10 m	L4	L3	L2	L1	Low trees	L			

Growth form							e.g.
Tall shrubs >2 m	S3	S2	S1	S	S+	Hummocky grasses t t	e.g. Triodia
Low shrubs >2 m	Z3	Z2	Z1	Z	Z+		
Hummocky grasses		H2	H1	H	H+	Herba-ceous plants g k y	e.g. Cyperaceae Chenopodiaceae
Tussocky grasses and graminoids	G4	G3	G2	G1	G	G+	Range of tussocky grass genera with or without graminoid genera y	Compositae
Forbs			F1	F	F+		

Littoral complex   grey mapping tone only
Intertidal mosaic of mangroves
Low shrub, herbaceous plants, bare salt flats.

Notes

Foliage cover   the proportion of ground that would be shaded by a stratum if sunshine came from directly overhead and no other strata were present.

Shrub   woody plant that is multistemmed from, or from near the ground

Hummock Grass   Mound like plant up to 1 m in height with repeatedly branched stems and long spiny leaves

Graminoid   a plant, grass-like in form, but not floristically of the family Gramineae (grasses).

Forb   A herbaceous plant – not a grass, or grass-like form.

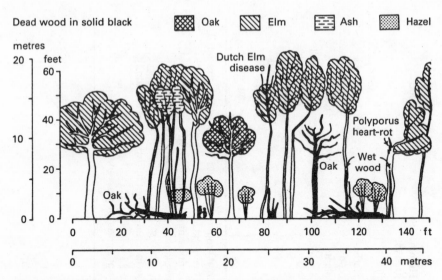

Dead wood in solid black ▨ Oak ▨ Elm ▤ Ash ▩ Hazel

Figure 4.6 Structure of Elm (Ulmus spp.) woodland in Hayley Wood, Cambridgeshire, UK. (Redrawn with kind permission from Rackham (1975).)

Note: 1 old spreading elms have given rise by suckering to the taller, younger, narrower, elm emergents;
2 the remains of large and small coppices of hazel (*Corylus*) and Ash (*Fraxinus*);
3 three common diseases of elm are illustrated.

gradient; for example, towards a lake margin, across a salt marsh or across a grassland–forest boundary.

2   Leisure woodlands. It has been suggested that beech (*Fagus sylvatica*) woodlands are the ideal type of natural habitat for recreation and therefore beech woods should be planted and encouraged near towns and cities.
Design a project to compare the stratification of beech woodland with other woodland types. Attempt to explain the differences observed. This may involve a survey of variation in light environment similar to that described in Section 2.7.

3   Vegetation stratification and bird populations. Locate a series of sites which form a sequence displaying increasing complexity of stratification. Using the Elton and Miller categories described in Section 7.4, count the number of different bird species seen in ten minutes in each of the stratification types. Explain any observed differences.

5 Floristic methods for describing vegetation

5.1 Floristic methods

Detailed surveys of vegetation often require data on species composition in addition to, or instead of, structural or physiognomic descriptions. Such studies involve the identification of individual species and also the assessment of abundance of species. The techniques applied are known as floristic methods of vegetation description. The collection of data at the species level is fundamental to biogeography and ecology. Individual plant species are the building blocks of plant communities (see Section 2.3) and satisfactory survey and explanation of the distribution of communities as well as individual species requires floristic description. Also detailed appreciation of plant-environmental relationships and the effects of management of vegetation by man can often only be fully understood by a floristic survey.

5.2 Quadrats

The usual sampling unit for describing the floristic composition of vegetation is the quadrat. Traditionally quadrats are square, although rectangular and even circular quadrats have been used. The purpose of the quadrat is to establish a standard sampling unit for describing the plant species composition of an area.

The conventional quadrat is illustrated in Figure 5.1. It has a metal or wooden frame and is often subdivided with lengths of wire or string. This subdivision increases accuracy in recording, since each sub-unit of the quadrat may be examined separately. The most common subdivisions are 1/10ths, 1/5ths or 1/4ths, of each side of the quadrat, giving 100, 25 or 16 sub-units respectively. Quadrats are easily made or purchased.

Long-term studies of vegetation change may require the establishment of permanent quadrats. Here the presence, absence or changing abundance of species may be recorded over a period of years. Corners of quadrats are usually

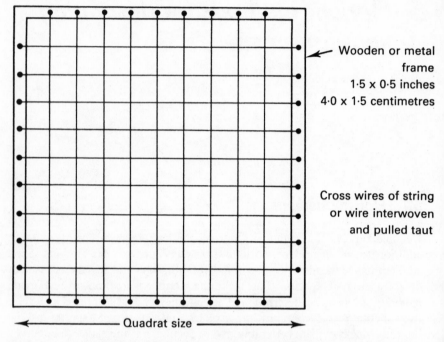

Figure 5.1   A conventional square quadrat with subdivisions

marked out with coloured pegs and are joined by string to define the edges of
the quadrat.

Quadrat size

The choice of quadrat size in designing a survey is very important, and will
vary from one vegetation type to another. The 1 m square quadrat is perhaps
most familiar but cannot be used in all habitats. Suggested quadrat sizes for
particular vegetation types of the British Isles are given in Table 5.1.

Table 5.1   *Suggested quadrat size for particular vegetation types in Britain*

Vegetation Type	Quadrat Size
Bryophyte and lichen communities	0.5 m × 0.5 m
Grasslands, dwarf heaths	1 m × 1 m − 2 m × 2 m
Shrubby heaths, tall herb and grassland communities	2 m × 2 m − 4 m × 4 m
Scrub, woodland shrubs	10 m × 10 m
Woodland canopies	20 m × 20 m − 50 m × 50 m (or use plotless sampling)

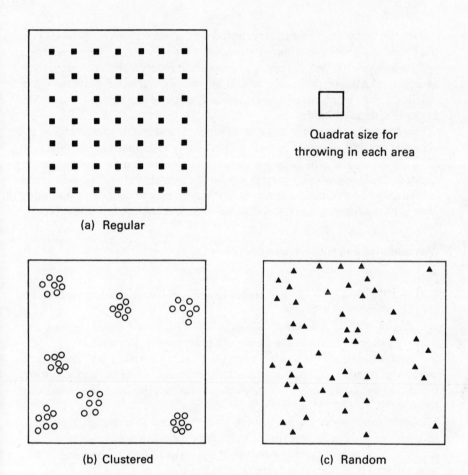

Figure 5.2 Pattern in vegetation: (a) regular pattern; (b) clustered pattern; (c) random pattern

Two groups of factors influence the choice of quadrat size to be employed. First, the size and morphology of the plant species under study and second, their pattern of distribution on the ground. While the significance of the size and shape of the plant species is readily apparent, the concept of pattern in vegetation may be less obvious. Pattern is defined according to whether individuals are distributed on the ground in a clustered, random or regular fashion, as exemplified in Figure 5.2.

Depending on quadrat size, different results will be obtained for clustered as opposed to regularly distributed species in a community, even though both types of species may have the same number of individuals present in the same area. In Figure 5.2 the regularly distributed species (a), has a much greater chance of being sampled if a quadrat of the size shown is thrown within the area, than another species with a clustered distribution (b), even though the number of individuals of each species present in each case is the same. The likelihood of a species distributed at random within the area (c) being recorded will lie between the two extremes. In practice, it is very difficult to accurately assess the degree of patterning within an area of vegetation without much detailed survey work. Many researchers acknowledge that this problem exists but take little account of it.

Smaller quadrat sizes are more easily studied, and enable much work to be achieved quickly. Larger quadrats will probably yield more reliable data on the vegetation, but will be more time-consuming to search. The optimum quadrat size is one which provides a reliable representation of the vegetation present, for the minimum amount of work. Attempts to define this optimum have been made using the concepts of minimal area and species-area curves.

Minimal area and species-area curves

Procedure
1   Find the middle of a representative area of the vegetation type to be studied.
2   Starting with the smallest possible quadrat size, usually containing just one or two plant species, count the number of plant species present.
3   Double the size of the quadrat using the pattern shown in Figure 5.3(a) and count the number of species present. Repeat the doubling and counting procedure until the number of species counted at each doubling of quadrat size levels off.
5   Cease work when further doubling of the quadrat either gives no new species, or a sudden increase in the number of species occurs.
6   Plot the graph of species numbers against quadrat size. The result is the minimal area or species-area curve (Figure 5.3(b)).

If a homogeneous area of vegetation is taken, then the curve of species numbers levels off and that point represents the minimal area necessary for sampling that community. The recommended quadrat size should then be slightly larger than the minimal area.

The method only works well if the vegetation being sampled is truly homogeneous, and is not continuously changing. If non-homogeneous units are taken, the species-area curve may level off within one locally homogeneous unit, before starting to rise again as the doubling of the sample size starts to sample a different community type. The selection of homogeneous areas also presupposes a certain amount of knowledge about the vegetation being studied, and often this information may not be available.

As an alternative to the determination of minimal area, the rough guidelines laid down in Table 5.1 may aid the choice of optimum sample size for particular vegetation types. However, they must not be taken as universally applicable.

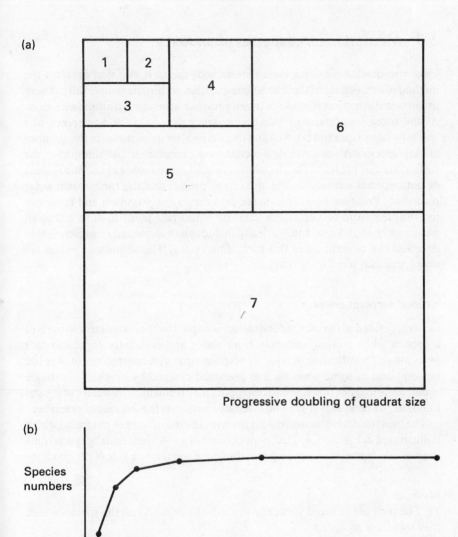

Figure 5.3 (a) Progressive doubling of quadrat area for minimal area and species-area curves; (b) The resulting species-area curve

5.3   *Measurement of species abundance*

Once the quadrat size has been determined, the next decision involves the method of assessment of the species composition within the sample unit. A very important distinction is made between presence/absence and abundance data. As the name suggests, with presence/absence data only the occurrence of a species within a quadrat is noted and there is no attempt to measure the quantity of each species. Abundance data include some measure of the amounts of the various species present. The decision as to whether to record abundance values or not depends entirely on the aim of the project and the time available for recording. Presence/absence data are extremely rapid to collect and represent the simplest form of vegetation data. An important point to note is that all measures of abundance automatically include the assessment of presence/absence and can be reduced to this form if necessary. The abundance values are supplying additional information.

Estimating plant cover

Cover is defined as the area of the quadrat occupied by the above ground parts of a species when viewed vertically from above and is usually expressed as a percentage. Stratification or layering of species may give cover values of over 100 per cent and in some areas such as grassland covered by bracken (*Pteridium aquilinum*) or scrub, this spatial overlapping may result in cover values of several hundred per cent. Cover may be estimated with varying degrees of precision.

The most objective measurement involves the use of a cover pin frame which is illustrated in Figure 5.4. The frame consists of a row of ten equally spaced pins in a wooden frame, with the length of the frame equal to one side of the quadrat.

Method
1   The pins are lowered vertically towards the ground and the species which they touch are recorded.
2   The procedure is repeated ten times at even intervals across the quadrat to give a total of 100 pin samples.

The pin diameter should be as small as possible, since pin size has been shown to be correlated with an exaggeration of cover values. If more than one species is touched as the pin is lowered, more than one hit is recorded, resulting in a total cover value for the quadrat of over 100 per cent. The method is time consuming, and may be affected by clustering or regularity in the vegetation. It is also difficult to use in tall or shrubby vegetation, such as bracken or mature scrub.

An alternative to the cover pin frame is the subjective assessment of cover. This involves a visual estimate of the percentage area of the quadrat occupied by each species. A number of recording scales have been devised to aid assessment. Some field workers record values in the complete range of 0–100 per cent, at

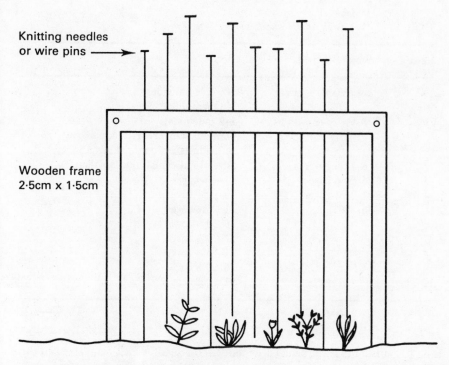

Knitting needles
or wire pins ⟶

Wooden frame
2·5cm x 1·5cm

Figure 5.4 A cover pin frame; 0.5 m² or 1 m² in size

1 per cent intervals. Others use the Domin or Braun-Blauquet scales. Their divisions are given in Table 5.2. A plant species which covers 55 per cent of the area of a quadrat would be recorded in the field as having a Braun-Blanquet cover value of 4, or a Domin cover value of 8. Clearly, greater precision is available with the Domin scale.

Since these methods are subjective, they are open to bias. Recorders may exaggerate the importance of attractive, conspicuous or readily identified species and underrate others. Nevertheless, either method is rapid and effective in its use. In grassland situations, where individuals of a particular grass species cannot be identified, the subjective estimation of cover is essential.

Frequency

The frequency of a species is defined as the chance of finding it in a particular sample area or quadrat. Frequency within an area is obtained by throwing a series of quadrats and expressing the number of quadrats containing each species as a percentage of the total number thrown. More commonly, a quadrat is subdivided into 10 × 10 sub units, as in Figure 5.1, and percentage frequency is determined by recording presence/absence of a species in each of

Table 5.2   *The Domin and Braun-Blanquet cover scales*

Value	Braun-Blanquet	Domin
+	Less than 1% cover	A single individual
		No measurable cover
1	1–5% cover	1–2 individuals
		No measurable cover
		Individuals with normal vigour
2	6–25% cover	Several individuals but less than 1%
3	26–50% cover	1–4% cover
4	51–75% cover	4–10% cover
5	76–100% cover	11–25% cover
6		26–33% cover
7		34–50% cover
8		51–75% cover
9		76–90% cover
10		91–100% cover

the 100 sub-units. Frequency is dependent on quadrat size, plant size and the patterning of the vegetation. A distinction is also made between root frequency and shoot frequency, which means that it must be clearly stated whether individuals recorded were rooted inside the quadrat or whether the aerial parts of plants rooted outside the quadrat but which may extend over into it, were also counted. This point is usually discussed with respect to frequency, but it should be noted that the same principle also applies more generally, and particularly to presence/absence data.

Density

Density is a count of the number of individuals of a species within a quadrat. Subdivided quadrats are essential for accuracy in counting and the method assumes that individuals of a species can be separated, which is a serious problem in grasslands and with shrubs where, for example, apparently distinct hedgerow elms may all be developing from the same rootstock (Figure 4.6). Density measurements are most frequently used in studies of herb species or saplings of trees and are rarely used in the description and analysis of whole communities. The method is time consuming to use and values derived are subject to the same limitations concerning the size of quadrats. Quadrat size must be kept constant when using density counts because the number of individuals found is entirely a function of the area examined.

5.4 Checksheets

The collection and subsequent analysis of floristic data will be facilitated by a pre-field work session considering:

1 what floristic data are actually required?
2 do all the species need to be identified or will more limited data be adequate?
3 the development of a field checksheet which will help first to recall the details of the data needed in each quadrat, and second by standardizing and simplifying actual recording. A typical checksheet, for a vegetation survey of part of the Dartmoor National Park in the South West of England, is illustrated in Table 5.3.

The species likely to be encountered are listed together with all the locational and environmental data necessary for interpretation. Many methods for the study of soil variables are described in Briggs (1977) and Smith and Atkinson (1975), while techniques for geomorphic and climatic factors are presented in Briggs and Smithson (1985).

The production of a checksheet also makes loss of data less likely. Where records are taken by different observers in field note books, part or sometimes even all of the data often mysteriously disappear down the 'data drain'. For beginners, the checksheet may also include details of the Domin and Braun-Blanquet schemes and/or simple identification keys to the species.

5.5 Spatial sampling

Regardless of whether physiognomic or floristic data are being collected, a choice of sampling design has to be made. Problems of sampling are frequently neglected in simple studies of vegetation, but may nevertheless be very important. Where an area is being studied, then one of three main approaches to sampling may be used.

1 Random sampling
2 Systematic sampling
3 Stratified sampling (random or systematic).

Random sampling

If analysis of species and or environmental data using inferential statistics (Chapters 11–13) is anticipated, then quadrats should ideally be located at random. A grid of co-ordinates is set up over the study area, and pairs of random numbers are taken to locate each quadrat (Figure 5.5(a)). A table of random numbers is presented in Table 5.4. A starting number may be taken from anywhere in the table, for example in the bottom right corner or the

Table 5.3   *A typical vegetation survey checksheet*
DARTMOOR–GUTTERTOR, VEGETATION PROJECT CHECKLIST.

Date *17 May 1984*   Surveyor *J. Merrydew*   Weather *previous - wet windy cold*
present - dry windy cold.

Quadrat No. *83*   Size *1m²* Location *below Gutter Tor*
(IF) Transect No. *N/A*   Grid Ref *SX 5806672*
Soil Sample No *83*   Aspect *NNE* ; Bearing *25° N mag.*
   Peat depth *180mm*   Slope angle *20*
   pH   *4.0*   Altitude *290 M.O.D.*

Field Soil Moisture Estimate
V. Wet;   *Moist* ;   *Fairly dry*;   *Dry*;   *Very dry*;
   ✓

Field Grazing Pressure Estimate   (IF) Age of Calluna *? 2 Years*
No. of faecal pellets per 1 m²/5 m × 5 m   Max. Veg. Height   *0.31m*
   Cattle   3
   Sheep   7
   Horses   *1*
   Rabbits   2

	Cover %	Domin value		Cover %	Domin value
Agrostis setacea (curtisii)	77	9	*Polygala* spp		
Agrostis tenuis (capillaris)	5	4	*Potentilla erecta*		
Anthoxanthum odoratum			*Pteridium aquilinum*		
Calluna vulgaris	+	1	*Sedum* spp		
Cladonia spp			*Sieglingia decumbens*		*(Danthonia decumbens)*
Erica tetralix			*Sphagnum* spp		
Erica cinerea			*Trichophorum* spp		
Eriophorum spp			*Ulex* spp		
Festuca ovina			*Vaccinium myrtillus*	+	+
Galium saxatile	+	1	Bryophytes	20	5
Juncus spp			Other species		
Luzula spp					
Molinia caerulea	5	4			
Nardus stricta	5	4			
Narthecium ossifragum					
Pedicularis palustris					

Table 5.3   cont.

Comments

Soil sample taken 0–50mm. Occasional burnt *Calluna* stems in adjacent area.

Domin scale

Class + occurring as single individual with reduced vigour: no measurable cover
    1 occurring as one or two individuals with normal vigour; no measurable cover
    2 occurring as several individuals; no measureable cover
    3 occurring as numerous individuals; but cover less than 4% total area
    4 cover up to 1/10 (4–10%) total area
    5 cover up to 1/5 (11–25%) total area
    6 cover up to 1/4 to 1/3 (26–33%) total area
    7 cover up to 1/3 to 1/2 (34–50%) total area
    8 cover up to 1/2 to 3/4 (51–75%) total area
    9 cover up to 3/4 to 9/10 (76–90%) total area
    10 cover up to 9/10 to complete (91–100%) total area

spp – indicates presence of specimens which cannot be attributed with certainty to one particular species of a genus; probably two or more species within the genus are involved

New species name changes are shown in brackets e.g. *Agrostis tenuis (capillaris)*

centre. Subsequent numbers must be taken in a systematic way, by working up or down a column, or along a row. With such methods, personal bias in the choice of sampling points is eliminated.

Where a grid is difficult to set up, for example in woodland, an alternative approach is to use a random walk procedure whereby a sample point is located by taking a random number between 0 and 360 to give a compass bearing, followed by another for a number of paces. Once this point has been reached and the vegetation has been described, subsequent samples are taken in exactly the same way.

Over larger areas, grids for random numbers may be set up on aerial photographs if available, although problems often occur in trying to exactly locate a point identified on an aerial photograph, when actually on the ground.

Systematic sampling

If the vegetation is sampled at fixed intervals, usually along a line, this is known as systematic sampling (Figure 5.5(b)). Such a systematic approach is often combined with the transect approach (see Section 5.6) where a sampling line is set up across areas where there are known to be clear environmental gradients. Care has to be taken with systematic sampling to ensure that the sampling interval does not coincide with some marked patterning in the vegetation. An example is where a meadow which has been subject to past agricultural practice is being studied and a form of ridge and furrow survives. If the fixed sampling interval is the same as the distance between two ridges, then a heavily biased sample will result.

Stratified sampling (random or systematic)

Stratified sampling is a particularly useful technique in vegetation work,

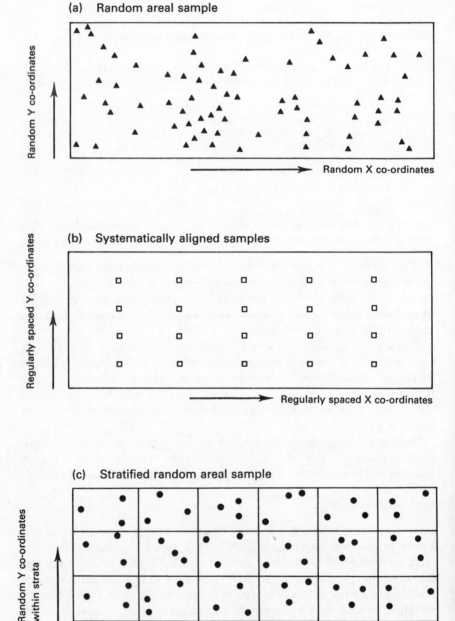

Figure 5.5 (a) A sampling grid comprising sampling points located using random number tables; (b) A systematic sampling grid; (c) A stratified sampling grid

Table 5.4 *A table of randomly generated numbers* (Reproduced with kind permission from R. A. Fisher and F. Yates (1974), *Statistical Tables for Biological, Agricultural and Medical Research*, 6th edn, Longman.)

Random numbers

03 47 43 73 86	36 96 47 36 61	46 98 63 71 62	33 26 16 80 45	60 11 14 10 95	53 74 23 99 67	61 32 28 69 84	94 62 67 86 24	98 33 41 19 95	47 53 53 38 09
97 74 24 67 62	42 81 14 57 20	42 53 32 37 32	27 07 36 07 51	24 51 79 89 73	63 38 06 86 54	99 00 65 26 94	02 82 90 23 07	79 62 67 80 60	75 91 12 81 19
16 76 62 27 66	56 50 26 71 07	32 90 79 78 53	13 55 38 58 59	88 97 54 14 10	35 30 58 21 46	06 72 17 10 94	25 21 31 75 96	49 28 24 00 49	55 65 79 78 07
12 56 85 99 26	96 96 68 27 31	05 03 72 93 15	57 12 10 14 21	88 26 49 81 76	63 43 36 82 69	65 51 18 37 88	61 38 44 12 45	32 92 85 88 65	54 34 81 85 35
55 59 56 35 64	38 54 82 46 22	31 62 43 09 90	06 18 44 32 53	23 83 01 30 30	98 25 37 55 26	01 91 82 81 46	74 71 12 94 97	24 02 71 37 07	03 92 18 66 75
16 22 77 94 39	49 54 43 54 82	17 37 93 23 78	87 35 20 96 43	84 26 34 91 64	02 63 21 17 69	71 50 80 89 56	38 15 70 11 48	43 40 45 86 98	00 83 26 91 03
84 42 17 53 31	57 24 55 06 88	77 04 74 47 67	21 76 33 50 25	83 92 12 06 76	64 55 22 21 82	48 22 28 06 00	61 54 13 43 91	82 78 12 23 29	06 66 24 12 27
63 01 63 78 59	16 95 55 67 19	98 10 50 71 75	12 86 73 58 07	44 39 52 38 79	85 07 26 13 89	01 10 07 82 04	59 63 69 36 03	69 11 15 83 80	13 29 54 19 28
33 21 12 34 29	78 64 56 07 82	52 42 07 44 38	15 51 00 13 42	99 66 02 79 54	58 54 16 24 15	51 54 44 82 00	90 18 48 13 26	38 18 65 18 97	88 72 13 49 21
57 60 86 32 44	09 47 27 96 54	49 17 46 09 62	90 52 84 77 27	08 02 73 43 28	34 85 27 84 87	61 48 64 56 26		37 70 15 42 57	65 65 80 39 07
18 18 07 92 46	44 17 16 58 09	79 83 86 19 62	06 76 50 03 10	55 23 64 05 05	03 92 18 27 46	57 99 16 96 56	30 33 72 85 22	84 64 38 56 98	99 01 30 98 64
26 62 38 97 75	84 16 07 44 99	83 11 46 32 24	20 14 85 88 45	10 93 72 88 71	62 95 30 27 59	37 75 41 66 48	98 97 80 61 45	23 53 04 01 63	45 76 08 64 27
23 42 40 64 74	82 97 77 77 81	07 45 32 14 08	32 98 94 07 72	93 85 79 10 75	08 45 93 15 22	60 21 75 46 91	98 77 27 85 42	28 88 61 08 84	69 62 03 42 73
52 36 28 19 95	50 92 26 11 97	00 56 76 31 38	80 22 02 53 53	86 60 42 04 53	07 08 55 18 40	45 44 75 13 90	24 94 96 61 02	57 55 66 83 15	73 42 37 11 61
34 85 34 35 12	84 35 09 50 08	30	42 40 07 96 88	33 85 29 48 39	15 89 95 66	51 10 19 34 88	15 84 97 19 75	12 76 39 43 78	64 03 91 08 25
70 29 17 12 13	40 33 20 38 26	13 89 51 03 74	17 76 37 13 04	07 74 21 19 30	72 84 71 14 35	19 11 58 49 26	50 11 17 17 76	86 31 57 20 18	95 60 78 46 75
56 62 18 37 35	96 83 50 87 75	97 12 25 93 47	70 33 24 03 54	97 77 46 44 80	88 78 28 16 84	13 52 54 94 53	75 45 69 30 96	73 89 65 70 31	99 17 43 48 76
99 49 57 22 77	88 42 95 45 72	16 64 36 16 00	04 43 18 66 79	94 77 24 21 90	45 17 75 65 57	28 01 19 72 12	71 96 12 82 96	69 86 10 25 91	10 50 71 75 12
16 08 15 04 72	33 27 14 34 09	45 59 34 68 49	12 72 07 34 45	77 25 50 25 83	92 12 06 76	99 66 02 79 54	69 36 38 25 39		48 03 45 15 22
31 16 93 32 43	50 27 89 87 19	20 15 37 00 49	52 85 66 60 44	38 68 88 11 80	97 49 29	24 02 94 08 63	48 54 69 28 23		
68 34 30 13 70	55 74 30 77 40	44 22 78 84 26	04 33 46 09 52	68 07 97 06 57	50 44 66 44 21	70 10 23 98 05	85 11 34 76 60	42 35 48 96 32	14 52 41 52 48
74 57 25 65 76	59 29 97 68 60	71 91 38 67 54	13 58 18 24 76	15 54 55 95 52	22 66 22 15 86	98 93 35 08 86	99 29 76 29 81	58 37 52 18 51	03 37 18 39 11
27 42 37 86 53	48 55 90 65 72	96 57 69 36 10	90 46 92 42 45	97 60 49 04 91	96 24 40 14 51	89 64 58 89 75	83 85 62 27 89	94 69 40 06 07	18 16 36 78 86
00 39 68 29 61	66 37 32 20 30	77 84 57 03 29	10 45 65 04 26	11 04 96 67 24	31 73 91 61 19	79 24 31 66 56	21 48 24 06 93	91 98 94 05 49	56 80 30 19 44
29 94 98 94 24	68 49 69 10 82	53 75 91 93 30	34 25 20 57 27	40 48 73 51 92	78 60 73 99 84	43 89 94 36 45	56 69 47 07 41	90 22 91 07 12	78 35 34 08 72
16 90 82 66 59	83 62 64 11 12	67 19 00 71 74	60 47 21 29 68	02 02 37 03 31	84 37 90 61 56	70 10 23 98 05	85 11 34 76 60	76 48 45 34 60	01 64 18 39 96
11 27 94 75 06	06 09 19 74 66	02 94 37 34 02	76 70 90 30 86	38 45 94 30 38	36 67 10 08 23	98 93 35 08 86	99 29 76 29 81	33 34 91 58 93	63 14 52 32 52
35 24 10 16 20	33 32 51 26 38	79 78 45 04 91	16 92 53 56 16	02 75 50 95 98	07 28 39 07 48	89 64 58 89 75	83 85 62 27 89	30 14 78 56 27	86 63 59 80 02
38 23 16 86 38	42 38 97 01 50	87 75 66 81 41	40 01 74 91 62	07 58 11 04 08	10 15 83 87 60	79 24 31 66 56	21 48 24 06 93	91 98 94 05 49	01 47 59 38 00
31 96 25 91 47	96 44 33 49 13	34 86 82 53 91	00 52 43 48 85	27 55 26 89 62	55 19 68 97 65	03 73 52 16 56	00 53 55 90 27	33 42 29 38 87	22 18 88 83 34
66 67 40 67 14	64 05 71 95 86	11 05 65 09 68	76 83 20 37 90	57 16 00 11 66	53 81 29 13 39	35 01 20 71 34	62 33 74 82 14	53 73 19 09 03	56 54 29 56 93
14 90 84 45 11	75 73 88 05 90	52 27 41 14 86	22 98 12 22 08	07 52 74 95 80	51 86 32 68 92	33 98 74 66 99	40 14 71 94 58	45 94 19 38 81	14 44 99 81 07
20 46 78 73 90	97 51 40 14 02	07 60 62 93 55	59 33 82 43 90	49 37 38 44 59	35 91 70 29 13	80 03 54 07 27	64 85 04 05 72	50 95 52 74 33	13 80 55 62 54
64 19 58 97 79	15 06 15 93 20	01 90 10 75 06	40 78 78 89 62	02 67 74 17 33	93 66 13 83 27	29 20 44 95 94	24 80 52 40 37	01 32 92 76 14	53 89 74 60 41
							28 34 96 53 84	48 14 52 98 94	56 07 93 89 30
05 26 93 70 60	22 35 85 15 13	92 03 51 59 77	59 56 78 06 83	52 91 05 70 74	02 96 08 45 65	13 05 00 41 84	93 07 54 72 59	21 45 57 09 77	19 48 56 27 44
07 97 10 88 23	09 98 42 99 64	61 71 62 99 15	06 51 29 16 93	58 97 77 09 51	49 83 43 48 35	82 88 33 69 96	72 36 04 19 76	47 45 15 18 60	82 11 08 95 97
68 71 86 85 85	54 87 66 47 54	73 32 08 11 12	44 95 92 63 16	29 56 24 29 48	84 60 71 62 46	40 80 81 30 37	34 39 23 05 38	25 15 35 71 30	99 82 93 24 98
26 99 61 65 53	58 37 78 80 70	42 10 50 67 42	32 17 55 85 74	15 29 39 39 43	77 69 10 61 78	44 91 14 88 47	89 23 30 63 15	53 34 20 47 89	43 11 71 99 31
14 65 52 68 75	87 59 36 22 41	26 78 63 06 55	13 08 27 01 50			71 73 26 95 62	87 00 22 58 40	92 54 01 75 25	
17 53 77 58 71	71 41 61 50 72	12 41 94 96 26	44 95 27 36 99	05 97 31 66 49	75 93 36 57 83	56 20 14 82 11	74 21 97 90 65	96 42 68 63 86	74 54 13 26 94
90 26 59 21 19	23 52 23 33 12	96 93 02 18 39	07 02 18 36 07	54 31 04 82 98	83 86 99 08 58	75 62 63 87 64	77 21 97 90 65	96 35 23 79 18	04 43 62 76 59
41 23 52 55 99	31 04 49 69 96	10 47 48 45 88	13 41 43 89 20	97 17 14 49 17	46 09 62 90 52	84 96 28 52 07	62 34 20 75	25 52 05 09	04 33 92 08 09
60 20 50 81 69	31 99 73 68 68	05 03 37 24 03	97 02 48 80 07	18 96 04 07 41	42 29 79 77 77	33 78 76 75 40	32 14 82 99 70	77 54 09 11 06	18 53 77 19 14
90 26 23 70 00	94 25 34 42 25	55 85 78 38 36	94 37 30 69 32	90 89 00 76 33	70 28 86 59		91 01 93 20 49	82 96 59 26 94	54 96 09 11 06
34 50 57 74 37	98 80 33 00 91	09 77 93 19 82	74 94 80 04 04	45 07 31 66 49	95 33 95 22 00	18 74 72 00 18	38 79 58 69 32	81 76 80 26 92	82 80 84 25 39
85 22 04 39 43	73 81 53 94 79	33 62 46 86 28	08 31 54 46 31	53 94 13 38 47	38 30 92 29 03	24 35 59 87 38	82 07 53 89 35	96 35 23 79 18	05 98 90 07 35
09 79 13 77 48	73 82 97 22 21	05 03 27 24 83	72 89 44 05 60	35 80 39 94 88	51 29 36 86 24	54 97 20 56 95	15 74 80 08 32	16 46 70 50 80	67 72 16 42 79
88 75 80 18 14	22 95 75 42 49	39 32 82 22 49	02 48 07 70 37	16 04 61 67 87	20 31 89 03 43	38 46 82 68 72	32 14 82 99 70	80 60 47 18 97	63 49 30 21 30
90 96 23 70 00	39 00 03 06 90			90 89 00 76 33	71 59 73 05 50	08 22 23 71 77	91 01 93 20 49	82 96 59 26 94	66 39 67 98 60

where areas of vegetation may first be divided into different but internally homogeneous types (the strata). Often such division will be based on physiognomic grounds such as areas of bracken within grassland or scrub within woodland. Sampling is then carried out either randomly or systematically within each vegetation type or stratum. In this case, major differences in plant communities are recognized before sampling commences. The systematic and random designs are combined in Figure 5.5(c). Here, using a grid for stratification, every grid cell is sampled, but the precise points investigated within each cell are chosen at random.

The number of samples which should be taken from a given area depends entirely on the nature of a particular project, together with the resources available for collection of data, which usually means the time and number of people available. However, for most short projects, a rule of thumb is that a minimum of between 25 and 30 points should be sampled.

5.6   Line and belt transects

Often, a preliminary study of vegetation distribution in the area being investigated will indicate a zonation in the presence/absence or abundance of particular plant species. Such changes of species numbers will reflect the varying tolerance of species to an environmental gradient (Chapter 2). Good examples of zonations are found:

1   on hillslopes, where slope form, soil moisture, and altitude combine to give a distinct sequence of landforms, soils and vegetation, known as a catena (Section 13.5);
2   across major changes in geology and lithology;
3   across natural transition zones joining marine, brackish, freshwater, and terrestrial habitats, (e.g. fens, salt marshes, sand dunes, lakes or canals);
4   across boundaries between vegetation types with clear physiognomic differences such as woodland margins, hedgerows, road verges or an invading bracken front;
5   on a global scale, across the biomes from the tropics to the polar latitudes, and from lower levels to mountain summits (altitude);
6   across gradients of human impact e.g. trampling, pollution, burning.

Where this environmental gradient can be readily recognized, then the transect approach to vegetation sampling is usually employed. Such transects may be defined as sampling lines set out across known environmental gradients. They are laid out in order to exploit the maximum variation in vegetation composition, and their location is thus clearly biased and non-random.

Line transects

This approach involves recording the presence or absence of species *continuously*, *systematically* or at *random* along a line marked by a tape or string.

Equipment
Graph paper; 30 m tape; string; two stakes.

Procedure
1 Identify the direction and nature of the environmental gradient.
2 Identify or label the species present.
3 Lay out a tape or string across the zonation.
4 Decide upon the sampling method. Three possibilities exist:
(a) Continuous sampling – presence and absence of all species recorded along the whole line.
(b) Systematic sampling – the species present or absent at fixed regular intervals along the line are recorded.
(c) Random sampling – the species present or absent at random points are recorded. Random number tables are used to decide the sampling intervals.
5 On graph paper, record the presence or absence of the species at the sampling points.
6 Indicate the presence of a species by a bar.
7 Add relevent environmental or land use detail.
8 The diagram may be extended to a form of profile diagram (Section 4.4) if required by sketching in the form of the plant.

An example of a line transect across the margin of a lake is shown in Figure 5.6. In many situations, the environmental gradient corresponds to a topographic gradient such as increasing water depth. In these situations, a levelled profile (Section 8.5) may effectively summarize and display the general character of the environmental gradient.

Belt transects

Belt transects are illustrated in Figures 5.7 and 5.8. They may be considered as a widening of the line transect or a continuous line of quadrats. Commonly, they are 0·5–1 m in width, although in woodlands and forests, the width may be much greater. All the types of abundance data described within quadrats may be collected as one moves down the belt transect. Thus, the survey method is particularly well suited to the investigation of the tolerance of species to environmental change and human impacts. Sampling may again be either systematic or random, depending on the aim of the project and the length of the transect. As with quadrat analysis, checksheets are essential to ensure consistency and accuracy of recording.

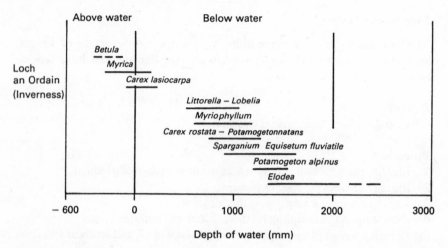

Figure 5.6   Zonation of emergent, floating leaved and submerged plant species in Loch an Ordain, Inverness, Scotland (Redrawn from Spence (1964); with the kind permission of Professor D. H. N. Spence.) (See also Plates 2.1 and 10.3 and Figures 8.3 and 10.2(a).)

An example of the use of a belt transect in the field of recreation ecology is shown in Figure 5.8. Here a gradient of impact from wheeled vehicles and walkers occurred along footpaths in the sand dune systems of the Lower Coorong, South Australia (Figure 4.5(b)). Using a quadrat size of 0·25 m × 0.25 m, a series of belt transects were laid out at right angles to the path or track. The species and environmental data collected from one transect are shown as a belt transect diagram in Figure 5.8.

Case study: Investigations on the effects of bracken invasion
Bracken (*Pteridium aquilinum*) is a fern which occurs widely over the globe and is seen as a weed or pest species in many situations. Extensive areas of pasture and heath in upland Britain and elsewhere have been invaded by this species, which has no feeding value, can be toxic to grazing animals and also suppresses the growth of more valuable forage species. Bracken reproduces both vegetatively and sexually but in large areas of upland Britain, sexual reproduction is limited by soil acidity and low temperatures. Thus, most reproduction occurs vegetatively through underground rhizomes which permeate the upper horizons of the soil. The leaves or fronds of the bracken develop from the surface rhizomes and because of these large underground storage organs, once they are established, the plants are very resistant to removal or damage, either by cutting, burning, trampling or extreme weather conditions.

The study of bracken distribution and invasion in relation to other upland pasture species in Britain is a useful project, and short belt transects across the edges of bracken areas may be used to study both the process of advance and

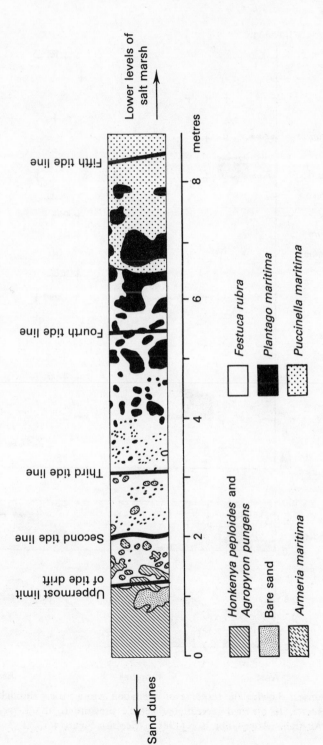

Figure 5.7 A plan of a belt transect showing the zonation of plant species at the upper edge of the salt marsh at St Cyrus, Kincardineshire, Scotland (Redrawn from Gimingham (1953); with kind permission of Professor C. H. Gimingham.)

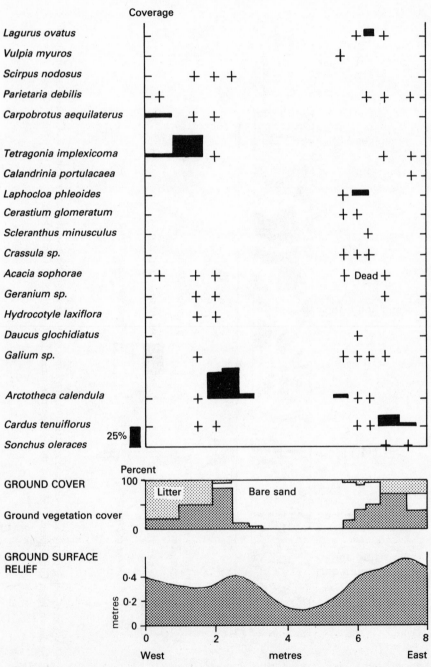

Figure 5.8   Belt transect showing the response of vegetation across a track through scrub (*Acacia sophorae*), to off-road recreational vehicle pressure in the Lower Coorong of South Australia (After Gilbertson (1983).) (See also Figure 4.5(a).)

the competitive mechanisms by which bracken suppresses the growth of other species.

In Figure 5.9, the results of a 4 m belt transect survey across the margin of a bracken-invaded area have been plotted as a series of lines representing variation in species abundance and vigour. The competitive effect of bracken and its ability to suppress other species is well illustrated in this diagram.

The mechanism by which bracken achieves this dominance is not just a function of its large growth form and dense foliage, which casts a dense shade and cuts down the light available to other species. Bracken also prevents growth of other species by the substantial quantities of litter which result from the die-back of each year's growth. In order to summarize the competitive effect of bracken, Mitchell (1973) derived an index of bracken interference, defined as:

$$\text{Index of bracken interference} = \frac{\text{Number of fronds} \times \text{Mean height}}{100}$$

Figure 5.10 shows a plot of the index against the weight of other species in the ground layer (gms). The data were derived from a belt transect. The effect of bracken in suppressing successful growth is clearly shown. An interesting aspect of this study is the derivation of a single index to represent the competitive advantage of a particular species along a belt transect. There are many other situations in which a similar approach might be adopted and such an index constructed, for example in a study of the vigour of *Ammophila arenaria* (Marram grass) across a sand dune system; the invasion of beach shingle and sand dunes by *Hippophae rhamnoides* (Sea buckthorn); or the invasion of woodlands by *Rhododendron ponticum* escapes from gardens or parks.

5.7 Plotless sampling

In many forest or woodland ecosystems, or in tall grassland, the capacity of conventional quadrat analysis may be considerably restricted by high densities of trees or impenetrable vegetation. The trees themselves may only be sampled by very large quadrats, which are likely to be difficult to install or locate. Several methods of sampling exist to overcome this practical problem, and are known as methods of plotless sampling. The simplest is the nearest individual method, which involves the location of random sampling points (Section 5.5) within a wood. The distance of each point to the nearest n individuals of a particular tree species is measured as shown for species A in Figure 5.11. Successive distance measurements to all species are then taken and the whole procedure is repeated for a series of random points. The density of each species is obtained from the formula:

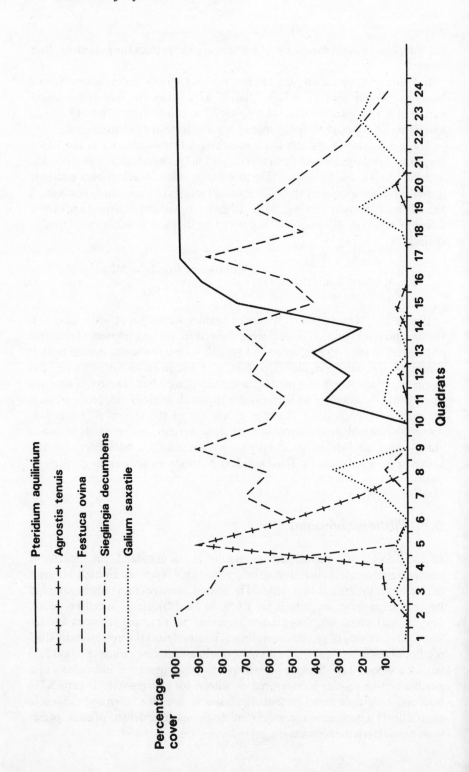

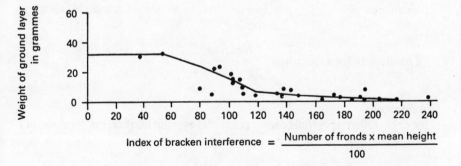

Figure 5.10 Variation in an index of bracken interference across a bracken (*Pteridium aquilinum*) invasion boundary. (Redrawn from Tivy (1973); with the kind permission of Oliver and Boyd.)

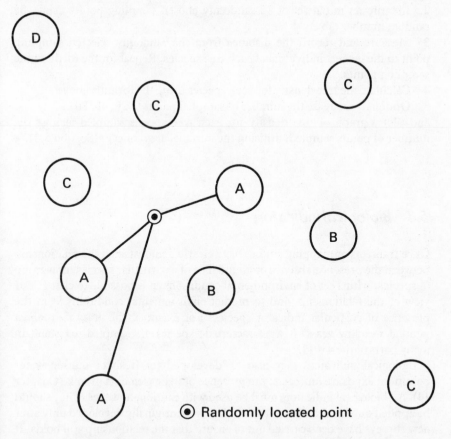

Figure 5.11 The nearest individual method of plotless sampling. A point is located at random and distances to the nearest individuals of each species are measured

Mean area $\quad\quad\quad\quad$ = (mean distance to nearest n individuals of a species)^2

Tree density for a species $= \sqrt{\dfrac{\text{Mean area}}{2}}$

Equipment
One or two 30 m tapes; four corner pegs; survey point marker; random number table (Table 5.4).

Method
1   In either an area of woodland, or in sparse vegetation typical of semi-arid regions, lay out the corners of a 30 m × 30 m, or 100 m × 100 m grid, as appropriate.
2   Identify an initial set of 10 randomly chosen sampling points using the random number table.
3   Measure and record the distance from the randomly selected sampling point to the nearest individual of each tree species. Repeat for the further nine sampling points.
4   Calculate the tree density for each species using the formula above.
5   Optional – increase the number of sampled points to 20, 40, 80 ...
and plot a graph of tree density for each tree species sampled against the number of points sampled, utilising the minimal area concept (Section 5.2).

5.8   Biological indicators

Once transects or mapping studies have clearly demonstrated the relationship between the presence/absence or abundance of a particular species, or a group of species, with a set of environmental conditions, it is possible to change our view of the relationship, and to predict environmental conditions from the presence of particular indicator species. For example, the grass *Poa annua* (annual meadow grass) is a characteristic species of trampled grassland in many parts of the world.

Biological indicators may also be developed for fresh or marine water pollution; exposure on coasts; the presence of toxic elements in soils (Chapter 14). All biological indicators must be used with caution, however. They should be applied outside the areas in which they were originally developed only after new surveys have been carried out to ensure that the relationship still holds. It is particularly important to establish that other factors do not distort or negate the relationships originally identified.

5.9 Projects

Classroom projects

Quadrat sampling and plant zonation

Using a transparent sheet, which has been marked out with a 1 mm grid, investigate the changing percentage cover value of the various species down the Scottish salt marsh shown in the belt transect of Figure 5.7. Do the results conform to your expectations based upon your knowledge of changing abundance of individuals of one species along an environmental gradient? 1 mm gridded transparent sheets may be purchased (see Appendix).

This exercise may be repeated using the vegetation and precipitation data for South Australia presented in Figure 4.5(a and b).

Spatial sampling strategies

You are in charge of organizing the ground survey to map the vegetation or habitats, occurring in a largely unexplored tropical island dominated by a central volcano with surrounding dense forest, whose foothills slope away into coastal plains fringed in mangroves. Aerial photographic cover is available (Plate 10.2), but reliable topographic maps are absent. Describe and justify a particular spatial sampling strategy your research team will follow. Indicate the theoretical and practical advantages and disadvantages of each spatial sampling strategy. Remember that the time and resources available for the survey will probably be restricted.

The efficiency of the sampling strategies

1 Transfer the spatial sampling designs and/or layouts shown in Figure 5.5 to a transparent overlay, and increase the number of sampling points by 25, 50 and 75, by either plotting new points on the overlay, or by moving (for example by rotation) the original overlay.

2 Plot a graph showing how the percentage of a particular vegetation type or habitat shown on a map (e.g. Figure 4.5(a)), alters as the size of the sample increases, for each of the three sampling strategies. Which method appears to settle down most quickly? Why is this method likely to be most satisfactory as a survey procedure?

Field projects

Minimal area curves in different vegetation types

Identify four or five different habitats which differ in both physiognomy and floristic composition. Apply the techniques of minimal area analysis described in Section 5.2 to each of the habitats and derive the species-area curves. Compare the results and explain the differences detected. The vegetation of the different areas must be uniform or homogeneous. Why is this so? Such areas are in reality, far less common than might be expected.

The floras of walls and paving stones
Using the principles of quadrats and sampling described in this chapter,
design a survey to study variation in the floristics of walls and/or pavements in
your home area. Many problems will occur and have to be solved. The survey
will require some standard height, sample length and orientation for studying
walls. The sample may need to be stratified according to wall or pavement
stone type. Aspects of biotic pressure such as trampling, dog impact and the
use of herbicides may need to be investigated in any study of pavements.

In the first survey, the identification of higher plants, often weed species will
be made. However, more adventurous students may wish to progress to
identification of mosses and lichens (for identification guides – see Appendix).
The statistical methods described in Chapters 11–13 will also be useful.

Recreation ecology and trampling of vegetation and soils
The study of the effects of recreation on vegetation and soils has become a
distinct sub-area of applied ecology in recent years. Using the information on
belt transects in Section 5.6 and Figure 5.8, design a project to study the effects
of recreation on vegetation and soils at a local beauty spot or tourist attraction.
Are there trampling resistant species? The effects of compaction of soils may
also be studied using methods for measuring bulk density (see Briggs, 1977)
and infiltration rate (see front cover, where students are measuring relative
variations in infiltration across a heavily used footpath by hammering in
open-ended tin cans of standard dimension into the ground and filling each
with a known volume of water). The time taken for this to drain or the volume
of water necessary to refill the can to a given level can be used as a relative
measure of infiltration rate. Hypotheses concerning relationships between
bulk density and infiltration may be generated and studied by using the
correlation and regression methods described in Chapter 13. Further aspects
of this type of study involving the use of trampling pins (Bayfield, 1971) are
described in Slatter (1978).

6 Rapid surveys of plant productivity

6.1 Introduction

Vegetation and ecosystems are dynamic in nature. Plants form the food source for all other organisms and all food webs are ultimately dependent on the amount of plant tissue or biomass available for consumption. The rate of growth of plants varies greatly in response to the whole range of environmental limitations, but most notably climatic factors of solar radiation and moisture. Changes in growth rate are affected seasonally and also by the stage in the life cycle of a particular plant species. The amount of plant tissue which is accumulated in a given area over a certain period of time is known as the primary production or productivity. An important related concept is that of biomass or standing crop, which is the amount of plant tissue in a given area at *one* point in time. Most methods for measuring primary production are based on the repetition of biomass measurements at several points in time, with any increase in biomass representing the net primary production.

Production studies are very important for a number of reasons. First, they can tell us a great deal about the dynamics of natural ecosystems. Second, they are of great value in agriculture and forestry, where man has domesticated certain crops for agriculture and the production of these crops and the factors affecting them become a very interesting applied problem. Equally, the measurement of forest production values to provide timber as a resource is an important area of research (Briggs, 1983). Third, as will be seen later in this chapter, concepts of primary production and its measurement can be useful in the study of plant–environment relationships through the application of bioassay techniques.

6.2 Measurement of primary productivity

Harvest methods

Biomass data are most easily obtained for short grasslands and herb fields by the harvest method. Different species may be either grouped or separated in the analysis.

Equipment
Shears; drying oven – 100°C for 24 hrs; metal or ceramic dishes; balance accurate to 0.1 or 0.01 gms.

Procedure
1   Clip the vegetation of a quadrat at ground level using shears or secateurs (the resulting crop may be sorted into species).
2   Determine, by weighing, the fresh weight of the crop of each species undried.
3   Dry at 105°C for 24 hours in an oven. Re-weigh. This gives the dry weight, with differences in water content of tissues eliminated.
4   The weight of dried plant material is the biomass.
5   Compare the fresh and dry weights and note the differing moisture contents of the species.
6   Check there has been no loss of biomass due to grazing or by death and shedding of plant parts since the previous biomass estimate.

Biomass estimates of trees and shrubs

Although studies of this type involving complete destruction have been made on whole trees, usually other methods of assessing biomass must be employed.

An index of tree size and biomass, commonly used in forestry work, is diameter at breast high (DBH). This is measured at 1.5 m up the trunk of the tree, either directly, using a specially calibrated pair of calipers, or by assuming that the tree is exactly circular in shape and working out diameter from the measurement of circumference. The diameter is then:

$$\text{Tree diameter} = \frac{\text{Tree circumference}}{3.14159}$$

Measurement of shrubs and saplings in woodland is very difficult. Subjective estimates of cover, combined with diameter measurement of the largest stems are two possible methods which may be employed.

Yield and productivity

If samples of biomass are taken from the same area of vegetation or animals at different points in time, any increase in the biomass figure between the first and second harvest is known as the yield or the net primary productivity of the community. Collection of biomass data is very time-consuming and with vegetation is non-repeatable since the sample is destroyed in the process. Care must be taken in the sampling of adjacent plots where the assumption is made that they are both exactly representative of the same community type. Careful thought also has to be given to the exact point of clipping at the ground surface, and as to whether root biomass should be included. No completely satisfactory method for assessing root biomass exists, although methods such as the washing of soil away from roots are used.

If the vegetation has been subject to grazing, then substantial amounts of biomass may have been lost. If this is serious, then an exclosure may be necessary to keep out large herbivores. Losses to smaller organisms are very difficult to measure and quantify. Loss due to death and shedding of plant parts may be estimated by placing litter trays under the vegetation where appropriate.

Assessment of performance involves the measurement of some relevant part of a plant which may provide an index of growth rate or vigour. Typical measurements are of leaf size; length and shape; plant height; flower or fruiting characteristics and tree/shrub annual growth ring increment.

6.3   Growth increment studies of trees and shrubs

Many species of trees and shrubs put on one annual growth increment each year, which can be recognized and measured when examining a section or a core from a tree. There are notable exceptions, for example in many tropical humid areas, there is insufficient seasonal variation to encourage the tree species to adopt this growth pattern. Occasionally, trees which normally put on one increment each year, when subject to severe stress from climate or disease, may miss a ring. On other occasions, two rings may be put on in a year. Analysis of tree growth annual increments offers a valuable means of investigating the changing performance of a species along an environmental gradient, e.g. soil characteristics, exposure or pollution. The method will also assess the effect on tree growth and productivity of various levels of forest clearance, soil erosion, fertilizer applications, drainage activities and other management practices such as coppicing. Growth increment studies may provide invaluable information on the population age structure of forests, the age of new landforms such as embankments, storm beaches or the moraines close to alpine or arctic glacier fronts.

Wood structure

A detailed knowledge of wood structure is required to establish the age of a tree or shrub from a study of its growth increments. As a general rule, in the British Isles, the oak (*Quercus*) is the most straight-forward tree to study. The ring structure of a young oak is shown in Figure 6.1. Cross sections through other common tree genera are shown in Figure 6.2. The identification of the age of a tree or shrub by growth increment studies is dependent upon that species producing 'annual' rings in its timber. This property is not present in all species.

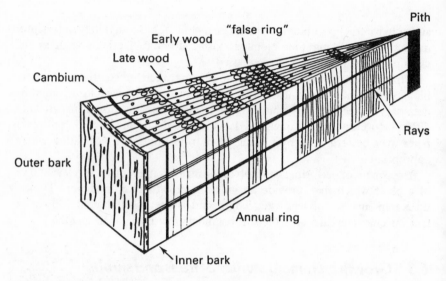

Figure 6.1 Diagrammatic section through a young oak trunk (Modified from Findlay (1975).)

Annual rings arise as a result of the activities of the cells in the vascular cambium which is shown in Figure 6.1. These cells divide to produce phloem or sieve tube cells on the outside edge. These phloem cells transport food to all parts of the roots or trunk. On their inside edge the cambium cells produce the tracheid cells of the xylem which transport moisture upwards from the roots to the leaves to facilitate photosynthesis. The character of xylem production varies at different times of the year. In spring, the supply of water is usually ample and the environmental conditions are generally favourable. At this time the cambium produces xylem cells that are thin-walled, with a large enclosed central space known as the lumen. In summer and autumn, conditions of mild stress appear as water supply decreases and heat and drought become a problem to the tree. The xylem's tracheid cells which form at this time have much thicker walls and enclose a smaller lumen. The end of the growing season is sharply delineated from the spring wood of the following year by the sudden transition to large, thin-walled cells.

In some conifers, especially pine species growing in semi-arid countries, times of severe stress may cause the tree to fail to put on a growth increment over large areas of its trunk in that year. On the other hand, particularly favourable conditions in late summer or autumn may cause large thin-walled xylem cells to develop and yield a 'false ring'.

Ring-porous and diffuse-porous trees

Hardwoods are divisible into diffuse-porous and ring-porous species as shown in Figure 6.2. The diffuse-porous tree species such as Birch (*Betula*), Hazel

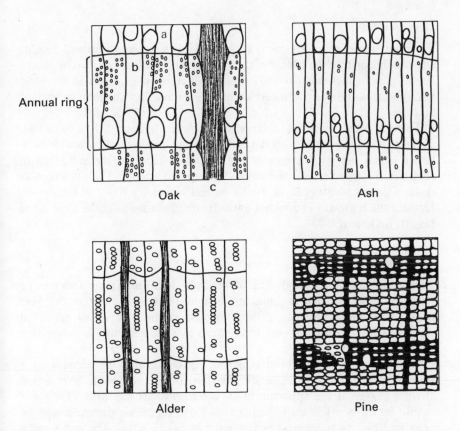

Figure 6.2 Sketch diagrams showing the structure of different types of wood. Each shows one complete annual ring. Oak and ash are ring-porous, while alder is diffuse-porous with scattered vessels and no distinct boundary to the annual ring. The structure of conifers, e.g. Pine, is very different – they have a much more homogeneous appearance with marked colour/density changes from early wood and late wood and some have large resin canals. Sketch diagram of oak; **a** large early wood vessel; **b** late wood; **c** a large ray. (After Morgan (1975).)

(*Corylus*), Sycamore (*Acer*) and Alder (*Alnus*) do not give rise to marked annual rings. There is an overall scatter of small vessels and a faint line of cells known as terminal parenchyma. These lines may delineate annual growth increments. However this cannot be relied upon.

The ring-porous group includes the important tree species of the oaks (*Quercus*), Ash (*Fraxinus*) and Elm (*Ulmus*). The two distinct zones of xylem cells or vessels are clearly marked and define each annual ring; the large, thin-walled tracheid cells forming the early spring wood. Obviously these species are the better trees for study.

Growth rates

The growth rate and productivity of tree and shrub species varies with height, age and throughout the course of the year as shown in Figure 6.3. The rate of lateral growth increment usually declines after rapid increases in the early years, because the tree has to develop more and more tissue as girth and height increase.

There is often no simple relationship between tree height, girth or age. Trees growing in relatively sheltered or favoured localities will probably have larger lateral growth increments per year than those of the same age in adjacent, but more exposed situations. In exposed woodland habitats such as those of the oakwoods of Black Tor Copse and Wistman's Wood on Dartmoor, Devon, much growth is directed towards lateral extension at the expense of height (Barkham, 1978).

Field studies

Large numbers of hardwoods and softwoods are felled each year. This may be for commercial reasons; as thinnings; because the trees are rotten; the trees might represent a hazard or they may be in the way of an approved development. Stumps or logs from these felled trees provide interesting material for study.

When permission has been obtained, softwoods may be cored with a Pressler corer as shown in Plate 6.1. Two horizontal cores should be obtained at right angles. If the opportunity to core hardwoods has arisen, the cores should be extracted in short sections (3–4 cm) otherwise the corer may be unextractable! All holes must be plugged with plastic wood. The core may be stored in a drinking straw.

The following data should be noted on site: sample codes for the cores; tree species; location, grid reference and site sketch; character of the soils, landforms, drainage, associated vegetation; recorder and date. The bark and inside end of the core must be clearly labelled.

In the laboratory, the cores are secured with white wood glue to a piece of grooved wood. The mount must be clearly labelled with the sample number and the bark and inside ends of the core should be marked with indelible ink or paint. When secure, the wood is trimmed with a very sharp knife and progressively sanded with sand paper of finer grades until the cell structure of the wood is clear. The wood and its rings may be examined with the very inexpensive and easy to use Beck hand microscope. This has a magnification of ×10 and has a graticule which permits the measurement of ring width. Suppliers of corers and microscopes are given in the Appendix.

Case study: The effects of limestone dust accumulation on the lateral growth of trees in south-western Virginia, USA.

Intensive quarrying of high-grade limestone may liberate large quantities of

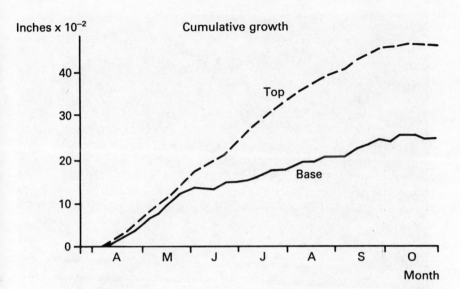

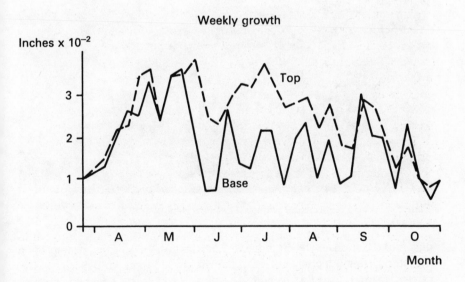

Figure 6.3 Cumulative growth and weekly rates of growth, as shown by stem diameter changes during the 1947 growing season measured 1.37 m above ground, and at the top (80 per cent of total height) as measured by growth bands, on *Pinus taeda* (Loblolly pine) (After Young and Kramer, (1952).)

Plate 6.1   The Pressler corer (Photo: D. Gilbertson)

dust into the atmosphere. Brandt and Rhoades (1973) identified its impact on
the lateral growth increment rate of four hardwood forest tree species – *Acer
rubrum* (maple), *Liriodendron tulipfera* (tulip tree), *Quercus prinus* and *Quercus
rubra* (oaks), at two sites. Both sites were similar except that one was in receipt
of large quantities of dust from quarries and processing plants, the second, a
control site, was not. Limestone quarrying started during 1945–8. Twenty
individual trees between 0.2 m and 0.3 m (DBH) were selected at random at
each site, and two increment borings obtained at right angles to each other.

Yearly growth increment data were plotted for each biennium i.e. measurements of the width of a couplet of two annual growth increments.

Analysis of the impact of limestone dust on the forest tree species requires two factors to be separated:

1 differences due to site factors other than limestone dust, e.g. soils, microclimate;
2 the impact of limestone dust.

The results (Figure 6.4) show slight differences in mean biennial growth rates of a species occurred between the sites before 1945–8. This may be identified by determining a growth ratio comparing the sites for the period before 1945 with the period after:

$$\text{Growth ratio} = \frac{\text{measurement of growth increment at subsequently dusty sites}}{\text{measurement of growth increment at control site}}$$

This analysis was then repeated for the years of dust accumulation, and the new ratio compared with the pre-1945 ratio. The results are given in Figure 6.4. The figure demonstrates that the maple (*Acer rubrum*), and the two species of oak (*Quercus rubra* and *prinus*) reduced their lateral growth rate significantly in the period following the initiation of limestone dust pollution. The tulip tree (*Liriodendron*) however prospered, increasing its lateral growth rate by some 76 per cent. The reasons for this need further investigation. Limestone dust in itself is likely to be deleterious to the photosynthetic and allied growth mechanisms of all trees. However, the success of the tulip tree may be due to the great decline in oak and maple growth causing a reduction in demand for soil, water and space resources by those species, resulting in less competition and a more beneficial habitat for the tulip tree.

6.4 Bioassay experiments

Bioassay experiments represent a very useful and important technique to the biologist and ecologist. Bioassay involves the germination and growth, usually under laboratory conditions, of a carefully selected plant species to demonstrate variations in plant production and performance. Numerous situations exist where bioassay experiments may be used successfully. Examples:

1 Carefully selected plant species may be used as indicators of general soil nutrient status or deficiency of specific nutrient elements. This approach may be valuable in comparing the general agricultural productivity of soils.
2 Pot experiments may be used to show the physical, chemical and biological limitations of growth media on disturbed sites. In reclamation of derelict land

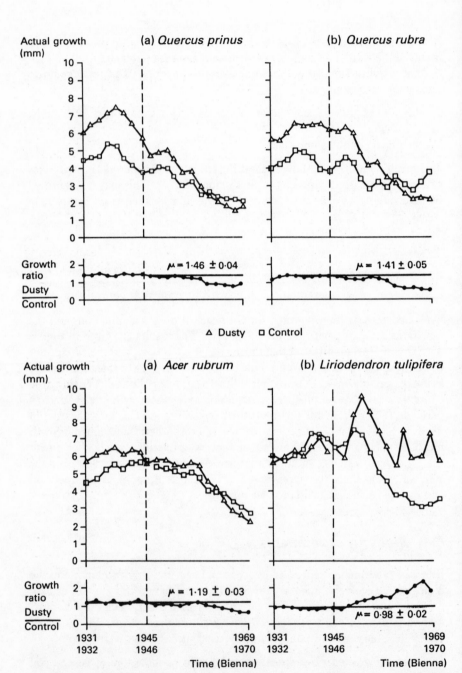

Figure 6.4   Analysis of lateral growth for four tree species on limestone soils in south western Virginia, USA; 1945–6 marks the period when quarry limestone dust entered the atmosphere in the area of the 'dusty' tree, but not of the more remote control trees (Redrawn with kind permission of Elsevier Applied Science Publishers Ltd, from Brandt and Rhoades (1973).)

there are frequently severe constraints to plant growth. Samples of toxic soil or spoil may be collected and the variations in performance of a species may be shown in bioassay experiments. Decisions may then be made as to action in reclamation schemes.

3 Pot experiments for samples in either of the above situations may also be used to study the potential effects of ameliorative treatments such as liming, fertilizing and correction of specific mineral deficiencies. Untreated soil or spoil samples may be compared with treated samples and variations in performance noted.

Setting up a bioassay experiment to study variations in soil productivity

Plant productivity is often very closely related to soil nutrient status. Thus variations of productivity and nutrient status may be studied using bioassay trials. Usually the experiments are carried out in pots and if, for example, the productivity of six soil types was being compared, samples of these six soil types would be brought from the field where their site conditions were fully described.

A good indicator plant species is then sown into the soils in the pots, either as seed or as young germinated tillers. These are then allowed to grow in standardized conditions for a given period, after which the plant material which has grown in each pot is harvested and the biomass or standing crop of the species is taken to represent the productivity of the various soils. Also species may show symptoms of particular stress such as yellowing or purpling of leaves or death of certain plant parts. Since the environmental conditions other than soil nutrient status are held constant, the yield of the species should reflect the relative nutrient status and plant growth properties of the original six soils.

General principles of bioassay experiments

1 The experiment must be set up in the context of some specific biogeographical or ecological problem as in the examples discussed above. They will often be used to test hypotheses (see Sections 1.3 and 11.2) derived from the study of distribution of a plant species in the field, assemblages of species in plant communities, effects of environmental controls, plant-soil relationships or the influence of management practices on plant species composition.

2 Experiments must always have a control. This means that some samples of the test species are grown in an ideal growth medium during the experiment and under the same laboratory conditions. In this way the effects of any external influences such as a bad batch of seed or an extreme environmental upset during the experiment can be seen and any variations in response due to factors other than those intended in the experiment noted.

3 Replicates must always be used. A particular soil type or combination of species being tested in a productivity experiment, for example, is known as a treatment. For each treatment, there must be several replicates. In the example of comparing six different soil types, at least three or four pots should be set up with exactly the same soil type, species and sowing density. At the end of the experiment, there should be a close similarity between the results in those pots and they may be averaged and any variations between them carefully noted. If there are very large discrepancies between replicates, the reliability of the experiment should be questioned.

4 The bioassay species must be carefully chosen. Again, this depends on the experiment. If plant growth is being used as a general index of soil conditions, then a species with suitable properties must be chosen. Usually such a species will have a relatively wide tolerance range, a fairly large biomass and growth form, a large seed, a rapid growth rate under experimental conditions and be easily harvested as an individual plant. Probably the most commonly used species in pot experiments in Britain is *Lolium perenne* (S23 perennial ryegrass). However, *Festuca ovina* (Sheeps fescue), *Festuca rubra* (Creeping fescue), *Agrostis tenuis* (Common bent) and *Holcus lanatus* (Yorkshire fog) are also often used. While grasses are the most common bioassay species, herbs, such as *Rumex acetosa* (Common sorrel), *Rumex acetosella* (Sheeps sorrel), *Potentilla erecta* (Tormentil) and *Galium saxatile* (Heath bedstraw) may be used.

5 Pots – normally pots 10–15 cm size are used. Again this depends entirely on the purpose of the experiment and on the growth form of the species.

6 Standardizing of conditions. Often, to speed up the results and to minimize variations in external environmental conditions, the experiment is set up in a plant growth chamber which may have artificially controlled temperature regimes and light conditions. This can be problematical where large experiments are established and it is quite feasible to set up experiments outdoors in summer months or indoors in a greenhouse in winter. However, if this is the case, pots should be moved around in the experiment, every week or so, using random numbers (see Section 5.5), to eliminate small scale environmental variations and possible competition effects between adjacent pots.

7 Harvesting and drying. Normally the tillers of grass or herb are harvested by clipping at the base of the stem where it enters the soil. It is important to standardize the position on the basal sheath of the bioasssay species where clipping occurs. Harvesting of roots may be attempted, but is usually very difficult. Once harvested, a careful check should be made that no soil particles have adhered to the plant as they will obviously bias the results.

Harvested samples for each pot are then put in separate unsealed envelopes and clearly marked with the details of the species, its growth medium and conditions. Drying is then carried out in an oven at 105°C for at least 24 hours. On removal from the oven and the envelope, the sample should be weighed to obtain the dry weight. This is the figure which is taken as a measure of plant productivity and performance. An accurate scientific balance is essential.

Case study: Bioassay trials as predictors of toxicities and nutrient status for reclamation of derelict colliery spoil heaps

In industrialized countries, revegetation of derelict and despoiled land resulting from mineral extraction and deep mining is a very important area of applied ecology. Of the many waste types which require reclamation, colliery spoil heaps are often the most intrusive in terms of landscape impact and are also difficult to reclaim. Colliery spoil is composed largely of shales which often become extremely toxic on weathering if they contain the mineral pyrites (FeS_2). This mineral weathers out in the following reaction with water and oxygen, to give various sulphur products, of which the most important is sulphuric acid (H_2SO_4).

$$2FeS_2 + 7O_2 + 2H_2O \rightarrow 2FeSO_4 + 2H_2SO_4 \text{ (Richardson, 1957)}$$

Clearly, the presence of sulphuric acid is detrimental to plants, but the shales are also often severely deficient in the macro-nutrients of nitrogen and phosphorus which are essential to plant growth.

Given sufficient time, the toxic pyrites will weather out and nutrient cycles will slowly be established. However, a recognizable soil profile and complete vegetation cover will not usually be formed for at least 100, if not more than 200 years.

Reclamation represents an attempt to speed up these processes and to establish a soil and vegetation cover over a very much shorter time span. An important aspect of reclamation is a site survey of variations in growth conditions prior to any landscaping work. Kent (1980, 1982) used bioassay trials to assist with prediction of plant growth problems in terms of both toxicities and nutrient deficiencies. The bioassay gives an indication of the likely productivity of the colliery spoil in its existing condition and hence the probable success of revegetation in the area from which the sample subjected to bioassay was taken.

A total of 34 tips were studied in the Barnsley and South Yorkshire areas of the National Coal Board. Tips were subdivided into smaller units on the basis of:

1 slope angle and form;
2 shale type;
3 existing semi-natural vegetation cover;
4 aspect.

Within each subdivision, four samples of colliery spoil were taken for bioassay testing. These were potted up in 10 cm pots and sown with the grass species *Lolium perenne* at 20 seeds per pot. Four replicates of each sample were used, and a set of four control pots was sown in a similar manner using Levington compost. Following harvesting, the grass tillers were dried at 105°C for at least 24 hours and the total dry weight of plant tissue in each pot was divided by the number of successful tillers to give a figure of mean weight/tiller for each pot. The percentage bioassay performance was then calculated from the formula:

$$\text{Percentage bioassay performance} = \frac{\text{Average weight/tiller for the four replicate samples}}{\text{Average weight/tiller for the four control samples}} \times 100.0$$

Since this procedure was carried out for each subdivision of the 34 tips, direct comparisons in terms of potential productivity and plant growth could be made. The spoils from the tips could be divided into three groups in terms of pH conditions and acidity resulting from the weathering of pyrites; those with a high acidity and low nutrient status resulting in low pH; those with low acidity and a high nutrient status giving high pH; and those in between. The mean scores for the percentage bioassay of site subdivisions in each of these three groups is shown in Figure 6.5, along with the mean scores for pH in each case. The relationship between pH as a measure of toxicity and nutrient status, and the bioassay performance (and hence productivity) is clearly shown.

In reality, interpretation of the results for individual site subdivisions in terms of reclamation problems was very much more complex and based on more detailed data. Nevertheless, the value of bioassay trials as an index of productivity and plant growth potential in this area of applied ecology is well demonstrated.

6.5   Classroom and field projects

Temporal variations in biomass and production

Procedure

1   Locate a relatively uniform area of grassland or lawn.

2   Set up a permanent quadrat of 4 m × 4 m using string and pegs.

3   Subdivide the quadrat into 16, 1 m × 1 m units.

4   For each week in the spring or early summer period, locate one of the 16 sub-quadrats using pairs of random numbers between 1 and 4, taken from Table 5.4.

5   Describe the floristics and abundance of the sub-quadrat using appropriate methods from Chapter 5.

6   Take a biomass estimate from the same sub-quadrat using the harvest technique described in Section 6.2.

7   Repeat this process for each of the other sub-quadrats at weekly intervals over the growth period.

8   Draw a diagram to show variation in floristics over the summer period.

9   Construct a scattergram to display the variation in total biomass from week to week.

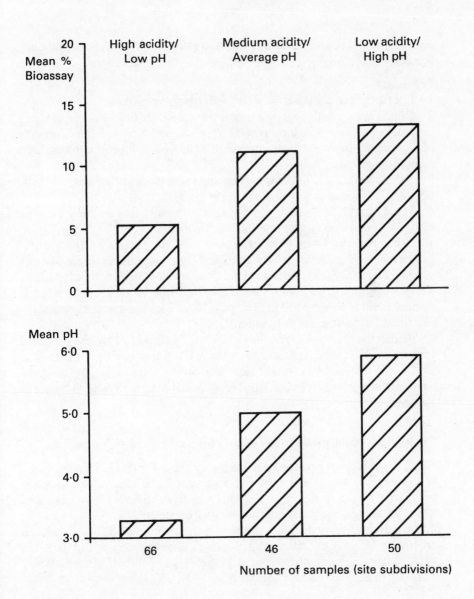

Figure 6.5 Percentage bioassay performance for three groups of colliery spoils classified by pH conditions

10 Draw another scattergram to demonstrate the change in primary production from week to week.
11 Comment on your results and think of factors such as variations in weather conditions (rainfall, cloudiness) which may correlate with variation in production.

Examination of the relationships between DBH height and growth rate in forest trees

Procedure

1   Locate an area of plantation woodland in your local area.

2   Obtain permission to investigate the trees and if possible to core dead trees or sawn logs using a Pressler corer (see Appendix). Also try to find out the date of planting of the wood. Bear in mind that the trees will probably have been planted as saplings.

3   Take a sample area of the woodland and using plotless sampling methods, describe the species and their density (Section 5.7).

4   For a sample of 30 trees of each species present (if possible), measure diameter breast high and tree height (see Figure 4.3).

5   Using recently sawn or dead timber and cut sections or from cores, determine the number of growth rings and hence an estimate of the age of a sample of trees of each species present. Also measure DBH at the same point.

6   Draw a scattergram to show the relationship between DBH and tree age. If a linear trend is apparent, it may be possible to use semi-averages regression (Section 13.4) to describe the relationship.

7   Predict the ages of the sample of 30 trees for each species present described in terms of DBH and height at stage 4, using the graph constructed at stage 6.

8   Discuss the primary production rate of the forest and any observed variations between species, or between the same species in different parts of the forest.

Comparison of the productivity of soils from differing environments

Using the guidelines presented for setting up a bioassay experiment described in Section 6.4, design and carry out a bioassay experiment to compare the productivity of a range of soils collected from different environments. Variation may be restricted in various ways, for example, soil samples may be taken just from varying rock types or from a range of different cultural environments, such as differing forms of agriculture or gardens and parks. In order to examine variation in these varying management environments, it may be useful to keep the geology constant.

7 Animal surveys

7.1 Introduction

There is a very large literature on the subjects of estimating the populations of animals in an area and the statistical techniques developed to assess the reliability of the estimates made. Methods of survey depend on:

1 the particular objectives of the study;
2 the particular circumstances and behaviour of the animal;
3 the nature of the habitat and terrain in the study area.

As a result animal surveys may be surprisingly simple or very, very difficult. In the past, many animal surveys have involved trapping wild vertebrates in remote locations, or insects or soil animals caught in various sticky, aromatic or light traps. In practice, important and interesting work can be carried out on species which are accessible, common, already known or easily identified and do not involve killing or trapping animals.

Animal surveys tend to be concerned with the estimation of one or more of the following:

1 the actual number of animals in area(s) at a certain time period;
2 the relative abundance of animals in area(s) at a certain time period;
3 the age distribution(s) of the population(s) and how this varies with location and through time;
4 the well-being i.e. the performance and condition of the animals;
5 the behaviour of the animals;
6 the impact of the animals upon the environment and vice versa;
7 the exploitation or conservation of the animals.

In many cases it is only possible to investigate a small sample of the total population, and it is necessary to establish the nature of the calibration factor which must be estimated in order to convert the data from the sample study into figures which reflect the characteristics of the entire population. In this study we concentrate on simple and effective methods of animal survey which require the minimum of equipment, few repeat surveys and minimal skills in identification. The main objectives here are to identify either the presence or absence of a particular species and to gain some idea of its comparative abundance from place to place and from time to time.

Interesting information on these subjects may be obtained by or from:

1   documentary and local government records;
2   casual or chance observations;
3   habitat classification and surveys;
4   examination of nests, tracks and signs such as droppings;
5   death assemblages and mortality data;
6   deliberate trapping and netting;
7   distribution studies and census counts by ground survey, aerial photographs, visual abundance scales and time-catch methods;
8   surveying preferred habitats;
9   capture-recapture methods.

All of these methods are aided by an understanding of the migration pattern, habitat preferences, home range or territory of the species and its changing use of the survey area through time. This can be illustrated by consideration of the behaviour of the red deer (*Cervus elaphus*). In upland Scotland this animal is an important resource for 'sport', i.e. hunting, as well as meat production. It also serves as a wildlife attraction. In other parts of the world the damage it causes to forest plantations makes it a major pest (Mitchell, Staines and Welch 1977).

The design of a survey of this species must take into account the following:

1   The red deer moves to the lower ground at night and occupies the higher ground during the day. It also occupies higher ground during the summer, descending to lower slopes and valley bottoms in winter.
2   There are often differences in the areas occupied by the males and females. In common with many wild ruminants, red deer live in groups with the mature males living apart from the females and young for most of the year. The males tend to occupy lower ground than the hinds in winter.
3   The home range of red deer varies according to the diversity of habitats in their area and other environmental and population pressures. On the island of Rhum in the Inner Hebrides the stags have an average home range of 800 ha, the hinds 400 ha. These areas are up to 16 times larger in north-east Scotland.
4   The stags may also travel up to 35 km from their traditional wintering or summer pastures to their favoured rutting locations.
5   The animals tend to seek shelter from the heat of the sun in summer and exposure and rain in winter.

Obviously they are most easily spotted against the snow covered slopes of winter. Consequently a survey with binoculars at this time, at 'artificial' feed sites in the valley bottoms might be one of the quicker, more comfortable and effective methods of surveying the deer population of an area.

7.2   Documentary, archive and local government records

Animals of economic importance which yielded revenue, or caused damage, will almost certainly have been documented in parish, estate or government records. For example, at the largest scale, Fenner (1971) has documented in detail the spread of the rabbit through Australia; at a regional scale, Mitchell *et al*. (1977) reviewed the status of the *c*. 270,000 red deer on Scottish moorlands; whereas at a local scale the parish records of some Yorkshire churches record payments for hunting and killing otters, which were supposedly consuming too many local salmon; while the game books of private estates provide useful information on the wildlife shot during hunts or for the master's table.

Local museums and reference libraries will help direct researchers to the best local data source.

The documentation of pests

Many local authorities have a statutory duty to maintain records of the presence and abundance of species considered to be pests. The evidence of presence may be direct, actual sighting or a catch; or indirect, such as damage of a characteristic type.

The house sparrow (*Passer domesticus*) or house mouse (*Mus musculus*) are commensal with man over much of the British Isles: i.e. the same habitat is occupied without any species having a biological effect on each other. Surprising gaps exist in our knowledge of this subject. For example, the distribution and abundance of the house mouse, for all its notoriety, has not been studied in depth until recently (Berry, 1970).

Data from the local pest/rodent control or environmental health officer may be of biological importance. The data may also yield interesting information on the differing perceptions and responses of different human communities. For example, in 1975, the Environmental Health Officer of the High Peak Borough Council (South Pennines, England), received 132 complaints of 'mouse infestation'. Some 89 per cent of the complaints came from urban areas; the remainder coming from rural areas. Several interesting possibilities emerge (Whiteley and Yalden 1976):

1   the house mouse is essentially an urban animal;
2   the rural human populations are less concerned with mice;
3   rural human communities are less helpless and more likely to take 'direct action' to control mice, e.g. rural households keep more cats to control mice.

This type of information makes it possible to generate interesting hypotheses on animals of medical or economic importance. Many government bodies collect such data on pests and also the movements of domestic and economic livestock through ports and airports.

7.3   Casual or chance observations

As the previous example shows, there is a considerable element of chance in the sightings of rodents. Estimates of the distribution and abundance of many animals rely on this type of casual or chance observation. Clearly the results obtained will be greatly influenced by the skill, perception and location of active recorders, who then forward their sightings to a biological records centre, for example, the Biological Records Centre, Institute of Terrestrial Ecology, Monks Wood Experimental Station, Abbots Ripton, Huntingdon.

The category of bird sighting known as 'accidental', where there are fewer than five records in a country for this century, is an extreme example of this type of observation and emphasizes the surprisingly large distances over which some animals will travel. The magnificent frigate bird (*Fregata magnificens*) from the South Pacific has this status in Great Britain, France and Denmark. There are at least 42 'accidental' bird species recorded for the British Isles during this century.

7.4   Habitat classification and surveys

The horizontal or vertical arrangement of vegetation (Section 4.4) provides a highly distinctive pattern of environments, resulting in a marked zonation in the type and abundance of the associated animal species. This again relates to the idea of preferred habitats.

The habitat description and classification system of Elton and Miller (1954) and Elton (1966) was devised as a rapid method of habitat survey for zoologists. It has subsequently been adapted for more general surveys of ecosystems and habitats for planning purposes and is also used in methods for ranking the ecological merits of habitats. The basic assumption of the method is that the structural complexity of vegetation, as represented by the degree of layering or stratification present, can be equated with habitat diversity, which in turn will encourage animal diversity.

Three major systems of habitat were originally defined, terrestrial, aquatic, and the aquatic-terrestrial transition. The terrestrial habitat is of the greatest interest with respect to vegetation and this is divided for survey purposes into four formation types on the basis of the height of the dominant plant species: open ground, field layer, scrub and woodland. These layers are illustrated in Figure 7.1.

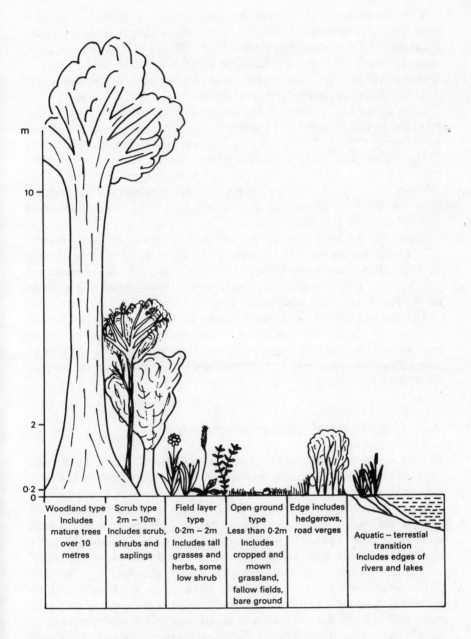

Woodland type	Scrub type	Field layer type	Open ground type	Edge includes	Aquatic – terrestial transition
Includes mature trees over 10 metres	2m – 10m Includes scrub, shrubs and saplings	0·2m – 2m Includes tall grasses and herbs, some low shrub	Less than 0·2m Includes cropped and mown grassland, fallow fields, bare ground	hedgerows, road verges	Includes edges of rivers and lakes

Figure 7.1 Categories employed in the habitat classification approach to animal mapping for conservation purposes developed by Elton and Miller (1954)

When combined with measurement of the area of each habitat type recognized, the technique provides a means of assessing not only vegetation structure but also ecosystem diversity. Where all four layers are present, as for example at the edge of a woodland, or in open-woodland habitat, species diversity will be high. In contrast, areas of open ground will have lower diversities of habitat and hence animal species.

Some animal species have distributions that are closely linked to particular plant species (see Section 12.1). In general, the greater the number of different plant species present within a stratum, the greater the variety of animal life. A further refinement of the method involves counting the number of plant species within a sample quadrat of each habitat layer recognized. The information on habitat structure, area of each type of habitat and plant species diversity, may then be combined in various ways to facilitate comparisons between differing vegetation types.

Linear features such as hedgerows and road verges (see Sections 11.4 and 15.5) may be counted as 'edge' types and measurements taken of their length between larger areas of vegetation, or their total length within standardized grid squares. Habitats adjacent to water bodies are assigned to a separate category, called the aquatic-terrestrial transition (Figure 7.1).

This method has the advantage of being very simple to apply and it is an efficient way of collecting information on animals through the proxy data on vegetation. The method is rapid in its application and is capable of modification and refinement in numerous ways, depending upon the purpose for which it is being used.

7.5   Death assemblages and mortality data

In 1974, only a few years after the late Donald Campbell was trying to break the world land speed record on the salt flats of Lake Eyre in South Australia, the same area was covered by a vast lake. The diversity and abundance of wildlife that had been able to utilize this huge area of water was indicated not only by the abundance of pelicans and other bird life, but also by the large quantities of plant and animal remains, especially fish, being washed up on the shores. Drift lines around lakes, rivers, marshes and beaches, as with 'oiled' seabirds, can provide important clues to the wildlife of an area.

The animal and plant remains found in this way, are known as death assemblages, brought together by waves, winds and currents. Species from many different habitats will have been mixed together and some remains may have been transported considerable distances. In other situations, the remains of identifiable vertebrates – sheep, rabbits, deer – will be found on moorlands, especially in late winter. The location of the remains and the size or age of the dead animal are data of considerable biological importance. Geologists and archaeologists view these remains as the first stage in the formation of the fossil record and are interested in the patterns of decay, scavenging and burial of the remains.

Man-made objects are providing similar death traps. For example shark nets off tropical or sub-tropical beaches, along with discarded cans may provide traps for animals. In many areas the domestic milk bottle, once overturned, has been used for actual survey (see Table 7.4). The car is proving an equally effective faunal sampler. Depending upon your location, examples of many animal groups such as insects, skunks, snakes, echidnas, kangaroos and hedgehogs may be found dead on the roadside.

Routine inspections of roadsides have provided good semi-quantitative survey data in Britain (Massey, 1972), New Zealand (Brockie, 1960) and North America (Haugen, 1944). Consideration of the survey methods shows the possible bias introduced by using these data to construct the species distribution.

Case study: Distribution and mortality patterns of the common hedgehog (*Erinaceous europaeus* L.)
The hedgehogs native to Britain and Europe have been introduced to many parts of the world. The advantages of studying them are that they are readily identifiable and that they are not migratory, but require suitable cover within a small area in which to feed, sleep, hibernate and rear their young. They will consume a wide variety of vegetable and animal foods and form a distinct hedgehog-rich zone around a favoured area of human settlement (Herter, 1965).

Data on hedgehog mortality along the roads around Scarborough, Yorkshire, England, have been collected by Massey (1972) and in North Island, New Zealand by Brockie (1960). The European hedgehog was introduced into South Island, New Zealand in the 1870s and to the North Island around 1910, where it is now widely distributed.

In the Scarborough example, counts were made and grid references recorded of all road corpses each month within a 24 km radius of the town over a five year period. In New Zealand, counts were made of dead hedgehogs on the main roads connecting Wellington, Napier, Taupo and Woodville over a period of one year.

The results of both surveys are shown in Figure 7.2. The British and New Zealand data reveal overall similarities in the variation of the mortality rate through a year. There are significant declines in mortality rates each winter, suggesting a hibernation period of approximately 4–5 months in each hemisphere. In Britain every year, the number of deaths rises abruptly to a peak in late spring or early summer, with a minor decrease before a second peak in autumn. These peaks coincide in time with increased activity associated with two peaks in breeding activity.

The Scarborough study also compared the death rates and distribution of corpses on straight roads, as opposed to winding or hilly roads; and on isolated

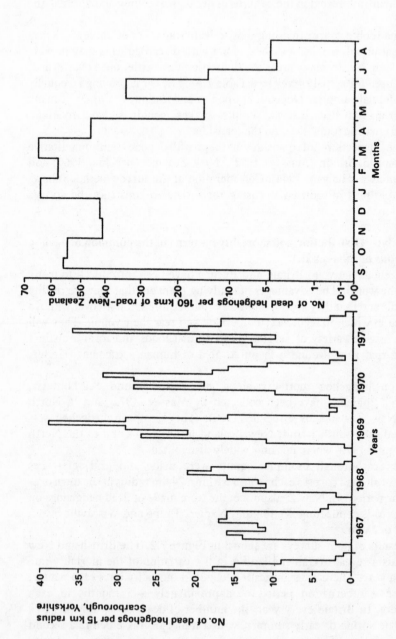

Figure 7.2 Variations in the road mortality rate for hedgehogs around Scarborough, England (Redrawn from Massey (1972)) and near Wellington, North Island, New Zealand (Redrawn from Brockie (1960)) Note the use of the log scale by Brockie.

roads as opposed to roads in more densely inhabitated areas. Most corpses were found near bends, the brows of hills, isolated farms, isolated buildings and on the outskirts of villages and towns.

7.6 Animal nests, tracks, signs and droppings

Animal tracks and signs provide good evidence for the presence of a particular species and may also provide information on both its patterns of movement and general activity. Such features are well known to native trackers and hunters. An excellent, cheap guide to the tracks and signs of mammals and birds in Europe, including the British Isles is provided by Bang and Dahlstrom (1974). With the aid of this type of interpretative guide it is possible to identify animal runs, routes, feeding signs, nests, prey remains, pellets, wallows and other features such as damaged trees which have been injured by a particular species.

Animal droppings

Pellets from many animals may yield valuable information on the prey they have consumed and hence the fauna within their hunting range. The droppings from two species of owl have been widely studied in the British Isles because of the relatively well-preserved state of animal bone and teeth found in their droppings. The species are the barn owl (*Tyto alba*) and the long-eared owl (*Asio otus*).

Faunal survey by the systematic collection and examination of the small mammal remains in such droppings has many advantages. First, it is relatively easy. One bird will use the same roost for long periods and will 'drop' at least two nightly pellets on the roost spot. Sometimes, nightly 'dropping' may continue for a period of several months. An identification guide to animal remains in owl droppings is given in Figure 7.3. More detailed, but still inexpensive guides, have been published by Glue (1969) and Yalden (1977). Other birds of prey eat flesh torn off their 'victims' and consequently their pellets lack the more easily identifiable, undigested skull remains.

Analysis of owl pellets

All animal droppings have the potential for transmitting disease. Always work in protective clothing: soak and dissect the droppings under Lysol to prevent infection.

Equipment

Shallow beaker, pair of mounting needles, soft paint brush, tweezers or

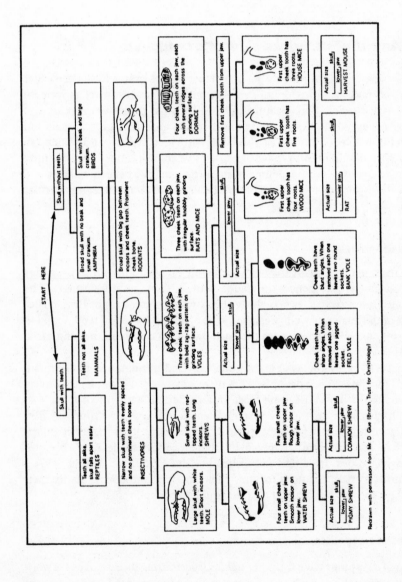

Figure 7.3 A simple visual scale for identifying small skulls in owl pellets (Copyright Mr D. Glue and Dr P. Morris; redrawn and reproduced with their kind permission.)

forceps, blotting paper, card, glue, identification guide (Figure 7.3).

Safety equipment
Face mask, rubber gloves, Lysol fluid.

Procedure
1 Wear rubber gloves and gauze face mask when collecting the droppings in the field. Use re-sealable polythene bags in the field and lab. Attach the sample label to the outside of the bag and seal inside a second polythene bag.
2 Immerse the material overnight in Lysol to prevent infection and leave to soak.
3 Dissect the pellet beneath the fluid surface.
4 Dry the skulls and the other bones found on blotting paper. Disinfect all equipment.
5 Mount the remains found on strong card using strong glue. Using the identification key, label and count the remains; make notes on the source of the droppings and other site details on the card.
6 Attempt to assess the numbers of individuals represented by the finds.

There are many other types of study which can be carried out on droppings. For example, the droppings of the larger vertebrates such as sheep, cattle or horses may represent important nutrient inputs to pasture. In the case of animals which defaecate in areas different from where they eat, a significant transfer of and relocation of nutrients may be effected.

7.7 Trapping and netting

The trapping and handling of animals is not to be encouraged except under the supervision of an expert, with due regard to risks of infection and to the well-being of the animal. Safe, useful data involving experience of trapping are available to students from two groups of animals – fish and molluscs.

Angling

Angling is a form of trapping. Many students have an extensive knowledge of fish taxonomy and ecology from the pursuit of their hobby. The problems of where to locate the 'trap' are immediately obvious. Problems of trap addiction or trap avoidance, in other words, the tendency of certain animals to keep taking the bait, or repeatedly avoid the bait, are readily investigated.

Studies of freshwater fish distributions will be aided by examination of Table 7.1, which presents a model showing the correlation between water conditions and the character of the fish population. In a few cases, the zones

Table 7.1   *Zones of British rivers characterized by particular fish species. These zones will not always represent a zonation down a river from upland streams to lowland rivers. Lowland rivers may have much greater discharges and velocities than many upland streams. In any river one or more of these characteristic zones may be missing*

	1 Trout zone	2 Grayling or Minnow zone	3 Barbel or Chub zone	4 Bream zone
Temperature range	less than 10°C	8–14°C	12–18°C	16–20°C
Water flow	rapid	swift	swift to smooth	slow
Water depth	shallow	less shallow	deeper	deep
River Bed	coarse gravel, stones, pebbles	stones, gravel sand	silt, mud	mud
Plant life	sparse	sparse	conspicuous	rich
Characteristic fishes	trout bullhead loach minnow	grayling minnow trout chub gudgeon dace barbel	barbel chub trout grayling	bream tench carp roach rudd eel pike perch

described may correspond with a progression from an upland stream to a lowland river. Excellent guides for the identification of European freshwater fish are provided by Maitland (1972) and Muus and Dahlstrom (1972).

7.8   Census and distribution studies

A census is a count of the population of animals in an area. Usually it is impossible to locate every member of the population and it is necessary to sample the population. Samples might be counted along belt transects or in quadrats following the principles described in Sections 5.2 and 5.6. The dimensions of the sampled area must represent a compromise between the size, frequency and mobility of the animal species studied and the skill and data sources available to the surveyor in that type of terrain. In many situations the larger vertebrates are readily studied with the aid of binoculars or oblique and vertical aerial photographs (Section 10.3). This is straightforward when dealing with partially, or wholly domesticated vertebrates such as sheep or horses on moorland pastures. It may be necessary to flush wild vertebrates from the cover of their habitats. Quantitative studies of wild vertebrates in such situations are not suitable as short term individual or class projects.

7.9 Visual-abundance scales and time-catch methods

In many situations the planner or reserve manager may not have the time or skilled workforce to survey adequately the numbers of individuals found in an area. Several survey methods have been developed to cope with this common problem.

Visual-abundance scales

Once the basic survey of an area has been completed, an appropriate visual-abundance scale may be devised. Table 7.2 shows a typical visual-abundance scale. In this case, the scale was designed to aid quantitative survey of molluscs on the inter-tidal rock platforms of south-west Britain. Visual-abundance scales are quicker and more readily repeatable than surveys using absolute counts and may be almost as accurate, especially on irregular rock surfaces.

Table 7.2 *A visual-abundance scale to aid rapid quantitative estimates of the molluscan faunas of inter-tidal rock platforms around south-west Britain* (After Crothers (1976).)

1. Acorn Barnacles and Small Winkles
7 More than 5 per sq cm
6 3–5 per sq cm
5 1–3 per sq cm
4 10–100 per sq decimetre
3 1–10 per sq dm: never more than 10 cms apart
2 1–100 per sq m: few within 10 cms of each other
1 Less than 1 per sq m

2. Limpets and Large Winkles
7 More than 200 per sq m
6 100–200 per sq m
5 50–100 per sq m
4 10–50 sq m
3 1–10 per sq m
2 1–10 per sq Decimetre
1 Less than 1 per sq Dm

3. Mussels and Piddocks
7 More than 80% cover
6 50–80% cover
5 20–50% cover
4 Large patches but less than 20% cover
3 Many scattered individuals and small patches
2 Scattered individuals, no patches
1 Less than 1 per sq m

4. Topshells and Dog-Whelk
7 More than 100 per sq m
6 50–100 per sq m
5 10–50 per sq m
4 1–10 per sq m: locally more
3 Less than 1 per sq m: locally more
2 Always less than 1 per sq m
1 Less than 1 per sq Decimetre

Time-catch methods

Animals that are much larger, more mobile and less frequently encountered may be studied using time-catch methods. Here abundance is determined by noting the frequency with which individuals are found per standard unit of searching time. It is also important to standardize the level of skill and effort put into the searching period. A typical abundance scale is illustrated in Table 7.3. It was developed by Öklund (1964) for his studies of the abundance of freshwater molluscs in an increasingly polluted lake – Lake Borrevan, south of Oslo, Norway. This scale is appropriate to the British Isles and it is easy to devise similar scales to cope with particular circumstances.

Comparisons between survey methods

Indications of the type and scale of the differences between the results of the different types of surveys may be gained by examination of Table 7.4. This displays the small mammal records obtained by a variety of survey methods in the English Peak District by Whiteley and Yalden (1976).

7.10   Sampling preferred habitats

Most animals are distinctly related to a particular type or types of habitat. Some species are dependent upon one particular species. For example the English cinnabar moth (*Tyria jacobaeae*) is restricted to ragwort (*Senecio jacobaea*) in its larval stage (see Section 12.2). It follows that when collecting data on animal distributions, it is important to possess a knowledge of habitat preferences of the species under study.

Table 7.3 *Scale of animal abundance based on time-catch methods as used in surveys of lacustrine molluscs (After Öklund (1964).)*

Scale of Abundance	Approximate number of individuals found per 30 minute survey period
1 Chance individuals	1–2
2 Small numbers	3–15
3 Numerous	16–30
4 Abundant	31–70
5 Mass occurrence	greater than 70

Table 7.4 *Comparison of the results of different techniques for recording small mammals in the English Peak District* (Data from Whiteley and Yalden (1976).)

Survey technique	*Sorex araneus* common shrew	*Sorex minutus* pigmy shrew	*Neomys fodiens* water shrew	*Clethrionomys glareolus* bank vole	*Microtus agrestis* field vole	*Apodemus* spp wood mouse	*Mus musculus* house mouse	Sample size
Traps	4.5	0.7	1.3	51.7	15.2	26.5	0.2	559
Owl pellets	19.4	6.7	1.5	3.0	56.1	12.9	0.5	2469
Bottles	59.8	6.4	1.7	10.7	7.6	13.9	0.0	1029
Casual observations	39.1	10.2	5.5	8.6	25.8	7.8	3.1	128

Percentages

Organizing a molluscan study

In order to demonstrate the principles of studying preferred habitats, the problems of studying molluscs are described in more detail.

Molluscs are useful organisms for study. They have low mobility; excellent cheap identification guides are available (Cameron and Redfern, 1976); (Evans, 1972); (Janus, 1965); (Kerney and Cameron, 1979). Detailed distribution studies have been published (Kerney, 1976). Once caught and studied the molluscs may be returned unharmed, to their original location. Molluscs are also found in a wide variety of habitats, including marine, brackish, freshwater and terrestrial environments. Their ecological requirements are well known (Boycott, 1934; 1936).

Some molluscan species can be expected to be found in a wide range of environments. Others will be much more restricted in their preferred habitat and geographical location. If time is limited, it is more effective to search the most favoured situations preferentially. In the British Isles these are chalk, limestone, old woodland or fen environments. In these locations, molluscan diversity is likely to be rich. Moorland, coniferous forest and acid grassland are likely to be poor in species.

Many land snails and slugs are susceptible to dessication. Therefore higher success rates will be experienced by searching in shaded and naturally moist locations such as beneath fallen tree trunks or under banks or in ditches. Exposed, dry environment have their own distinctive faunas. The time of the searching period is important. A lengthy period of dry weather will tend to drive most of the slugs a long way under ground. Damp conditions, especially at night in autumn, are most favourable.

The location of the larger specimens over 4–5 mm in length, requires close scrutiny of the ground surface. Smaller species or juveniles (1–2 mm in diameter) are even more difficult to see against a background of moss and leaves. In this case, either the leaf litter or the stems of herbs should be shaken through a nest of sieves. Sieves may be made, at little cost, using Nybolt cloth of 0·5 mm mesh size, stretched across a 20 cm × 20 cm wooden frame (see Appendix).

If only presence or absence data are required, the searching of preferred habitats with the aid of sieves is a satisfactory survey method. However, if quantitative comparisons of species abundance or population numbers are required, several aspects of the survey must be standardized. A field sample size of 0.2 m × 0.2 m, carefully sieved and searched, is now in use for basic survey and this may be seen as the equivalent of the 1 m × 1 m quadrat in floristic studies. These quadrats must be located in a standard manner which permits reliable interpretation. The discussion of sampling in Section 5.5 is applicable here. Finally it is necessary to standardize the amount of time, skill and effort spent searching particular areas. In this case, time-catch survey methods or visual abundance scales may be useful.

The association of particular species with new types of habitats or localities beyond their known range is likely to be noticed, in the first instance, as a result of casual or chance observations. In some cases where climate is thought to be a limiting factor in controlling molluscan distribution, it may be found that human activities may produce and sustain occurrences outside the main distribution because of locally-induced amelioration of the environment.

One further point needs stressing. Molluscs may be found alive or dead in habitats well removed from those with which they are normally associated. Two types of explanation may be involved. First the habitat properties of a location may mimic those of a more typical situation. For example species commonly associated with woodland may also occur in the similar, shaded, moist, litter and debris-rich environment within scree. Second, molluscs, especially aquatic species, are frequently transported on the feet or in the guts of wading birds and other vertebrates. Consequently, it is not uncommon to find large numbers of shells of aquatic molluscan species on dryland perching locations. These types of movement of molluscs and small creatures are important to palaeoecologists and geologists who endeavour to reconstruct past environments from fossil or sub-fossil remains.

7.11   Capture-recapture methods

The population of a species in an area may also be calculated by using the Lincoln index method of mark, release and recapture. This simple procedure is especially useful for animal groups such as molluscs or woodlice which are easily spotted and marked.

Equipment
Cellulose paint (not a bright distinctive colour); fine paint brush.

Procedure
1 In the study area – a garden, small copse etc., locate as many snails as possible in a standard period of time *c*. 30 minutes;
2 Count the numbers of individuals and apply a small blob of paint to the back of each specimen, allow to dry, and release, leaving them to mingle with the rest of the population in the area;
3 Return to the site after a period of 2–3 hours, or a day or so, collect as many individuals as possible in the same area in the same time;
4 Count the number of marked and unmarked specimens in this second sample.

The population may then be calculated using the Lincoln index formula:

$$\text{population number} = \frac{\text{no. marked on day 1} \times \text{no. captured on day 2}}{\text{no. of marked animals caught on day 2}}$$

The survey may be improved by counting the numbers of each species present when skills in identification develop, each species may be labelled with a different colour paint.

The method requires the following conditions to be satisfied:

1 that the population must have a finite boundary;
2 that the population should not alter as a result of births or deaths, or immigration or emigration to or from the population in the study area, in the study period;
3 marking must not alter the behaviour of the animals, the probability that they will be preferentially selected by a predator species; or their chance of recapture.

All of these conditions are unlikely to be fully satisfied in every study. Nevertheless it is possible to make a good approximation to them.

7.12 Reconstructing the population age structure

The management and conservation of animal communities requires more information than data on the number of individuals in that population. For example, the implications of a population dominated by old individuals, past breeding age, are different from those of a population dominated by young individuals. Consequently demographic information on the population

Cardium edule

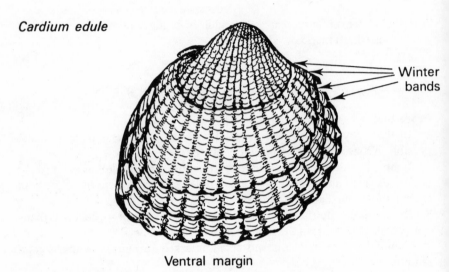

Winter
bands

Ventral margin

Figure 7.4   Winter bands and annual growth increments shown on the shell of the common cockle (*Cardium edule*) (After Boyden (1972).)

age-structure, birth rate, death rate and general well-being of the community are required.

Unfortunately it is often difficult to estimate the age of an individual. In vertebrates this may entail the examination of dentine layers of teeth or of the bone structure: practices which must be carried out after the death of the animal and require skill and supervision. In molluscan species it may be possible to define size-groupings, rather than age-groupings. Some species, such as the inter-tidal cockle *Cardium edule* display a marked reduction or cessation of shell growth in winter, as illustrated in Figure 7.4. This provides a simple means of ageing the individual. However, on occasion, severe disturbance of the mollusc by a storm or during handling will cause it to cease, temporarily, secreting its shell at its margin and yield a 'false band'. Such banding is not unusual in estuarine bivalve species.

Information on age-structure can be analysed graphically by using a type of histogram, which is commonly known as a fir tree diagram or population age-pyramid (Figure 7.5). This diagram has been constructed from information given in Table 7.5 on the population of red deer in Glen Feshie in Scotland. In general a wide base overlain by a rapidly contracting pyramid indicates a high birth rate and low survivorship. A broader peak to the histogram suggests a higher survivorship. Indentations in the pyramid walls indicate a relatively low population in an age group. The indentations may be especially interesting and might suggest earlier low birth rates or high infant mortalities; disease or exploitation. In the case of deer, hunting is a major factor in determining the shape of the population age pyramid. In contemporary human populations such troughs may be related to wars, economic or

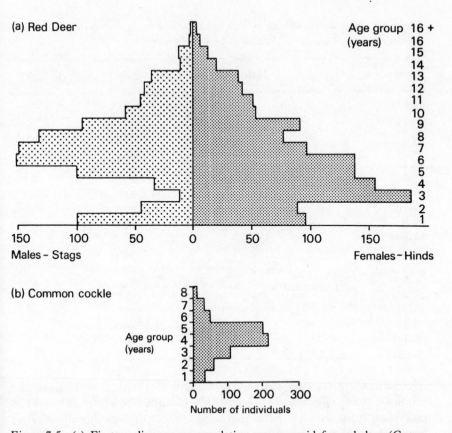

Figure 7.5 (a) Fir tree diagram or population age pyramid for red deer (*Cervus elaphus*) 1966–1973, Glen Feshie, Scotland. (After Mitchell *et al.*, 1977.) The diagram shows the selective impacts of hunting on the population. (b) Age structure of the common cockle (*Cardium edule*) in the Crouch estuary, Essex, England in 1968 (After Boyden (1972).) The small number of individuals in the oldest category reflects the high mortality of spat in the exceptionally cold winter of 1962–3; this is in contrast to the successful settlement of spat in the 1963–4 period. The data indicate the magnitude of natural fluctuations that can occur in the population density of the cockle

natural disasters on the one hand, or the impacts of affluence, new methods and attitudes to birth control on the other. Population age pyramids may be drawn up to show the actual numbers of males and females in each age group, or to express these data in the form of percentage frequencies.

A development of this technique is to construct life tables. This analysis requires data on the number of individuals alive and dying within each age-category. A good introduction is provided in Ricklefs (1979).

Well-being

The well-being of an animal is difficult to define and determine. It is often

Table 7.5   *Number of red deer in each age class shot in the period 1966–73 in Glen Feshie, Scotland* (Data from Mitchell *et al.* (1977).)

Age group in years	Males (stags)	Females (hinds)
16+	3	3
15–16	3	5
14–15	12	12
13–14	11	19
12–13	36	38
11–12	43	41
10–11	45	51
9–10	59	52
8–9	95	90
7–8	133	75
6–7	150	96
5–6	151	138
4–5	100	138
3–4	33	156
2–3	12	187
1–2	45	89
0–1	99	96

discussed in terms of 'performance' and 'condition'. Performance relates to the productivity of an individual or a population. Condition concerns the state of an individual at a given time. Performance covers growth, breeding and survival whereas condition relates to life expectancy or health (see Caughley 1977).

For most wild vertebrates performance is difficult to quantify in the field. Simple measures of performance such as skeletal size, body weight, reproductive status, or frequency of pregnancy, all need to be related to the age of the individual and normal annual cycles (see Table 7.6) before they can be reliably interpreted.

In agriculturally important vertebrates, condition is often assessed by chemical analyses of fats and/or other soft tissues of the animal. A more practicable approach for most students is to observe the animals at the time of year when food is in shortest supply and to follow the lead of Riney (1960) and develop a subjective visual scoring system, on a scale 1–10, which records the relative thinness of the individuals or the prominence of their ribs or pelvic bones.

7.13   Problems caused by the need for rapid faunal surveys

In many countries there is a legal requirement upon the potential developer of an area to provide a full description of the ecological and environmental importance

Table 7.6 *Weight and condition for four classes of red deer over the course of one year, Glen Feshie, Scotland* (Data from Mitchell *et al.*, 1977.)

Class	Item	early July	late Sept	late Oct.	late Nov.	late Feb.	late March	late April, early May
Stags	Carcase weight (kg)	77	92	82	78	68	65	66
(5–10 years)	KFI	2.0	5.6	1.6	1.5	1.4	1.3	1.3
	N	10	14	13	12	12	11	9
Non-lactating	Carcase weight (kg)	52	55	61	62	56	57	56
hinds	KFI	1.7	4.0	3.7	4.7	3.4	2.7	1.3
(5–10 years)	N	9	9	9	8	12	11	9
Lactating	Carcase weight (kg)	50	51	50	49	47	46	49
hinds	KFI	1.2	1.6	1.4	1.7	1.4	1.4	1.3
(5–10 years)	N	11	11	10	16	10	11	10
Calves	Carcase weight (kg)	16	24	26	24	25	26	28
(sexes pooled)	KFI	?	1.2	1.3	1.5	1.2	1.3	1.2
	N	9	10	12	14	12	11	11

Notes: carcase weight = weight of complete animal less alimentary tract. KFI = Kidney-fat-index.

These data need to be interpreted with the aid of information on the biology of red deer. In general wild red deer hinds have one calf per year. Puberty is usually reached in the age groups 2–3 or 3–4 years. The breeding season is well-defined. The mating season is from September to November, with a peak in October. Pregnancy extends over the winter and calving is in late May–late June in Scotland. Lactation is prolonged and may last until shortly before the following calving season. These trends are evident in the data above. In a longer time perspective, the pattern of body weight growth in red deer is typical of many large mammals. The most active period is in the early years up to 5 years, with a maximum body size and condition achieved between 5–10 years. There is a progressive decline above the age of 10 years. In general, the maximum body weight in hinds is attained at 6–8 years and 7–9 years in stags. In hinds the maximum condition appears to be attained approximately 2 years earlier than maximum body size or weight in stags

of the site and to assess the nature of the intended development in these terms. Environmental Impact Statements or legislation are not formally required in the British Isles, although a number of potential developers do provide them. Studies of this type frequently have time for only one site survey of the fauna. However, faunal studies must take account of the mobility of species and the fact that many populations experience natural fluctuations in the numbers of individuals present. Consequently unexpected absence or surprising lack of abundance might be the result of one or many factors; for example, direct avoidance of the surveyor, a nocturnal habit, a daily migration pattern from home to feeding area, seasonal migrations, or other human activities. As a result, a reliable view of the faunal abundance and productivity often requires surveys to be repeated over a period of days, months or years, depending upon the habits of a given species.

7.14 Projects

Classroom projects

Surveys of animal pests

1 Using a detailed map of your area, your own observations or data obtained from your local authority pest control officer, dog catcher's office, or environmental health department, attempt to explain the distribution of the following animals in urban or rural areas; the domestic rat, the house mouse, hedgehogs, dogs and cats, the common pigeon, starling, house sparrow, lice, snakes, poisonous spiders (overseas). Specialist pest control firms have a host of data on thoroughly unpleasant commensal creatures.

2 Compare and contrast the types of population data that are derived from direct counts, time-catch methods and visual abundance scales.

3 Apart from species lists what other types of information may be obtainable from repeated surveys of owl pellets from one site during the course of one year?

4 Describe and try to account for the differences in (a) condition and (b) population age structure between male and female red deer shown in Table 7.6. Background data on the biology of the red deer is given in Table 7.6 and Mitchell *et al.* (1977). What factors might be responsible for the differences between the population age structures of male and female red deer shown in Figure 7.5?

5 Construct and account for the shape of the population age pyramid for the people in your class or college. Include the teaching staff!

Classroom and field projects

Hedgehogs

Herter (1965) suggested that the greater use of agricultural chemicals in the open countryside may have caused an accelerated loss of hedgehogs from less populated areas.

1 List and justify other factors which would explain the differences in hedgehog mortality rates (a) in open countryside and inhabited areas; (b) between straight as opposed to bendy or hilly routes. See Davies (1957) for a formal definition of an index of hedgehog mortality frequency, or devise one of your own.

2 Devise and carry out a survey to assess the significance of seasonal variations in road traffic flows, in determining the observed temporal pattern in hedgehog mortality rates.

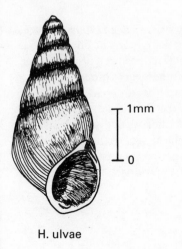

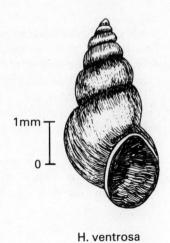

H. ulvae

H. ventrosa

Figure 7.6   The inter-tidal gastropods *Hydrobia ulvae* and *Hydrobia ventrosa* (adults)

3   To what extent will sampling in only fine weather affect our understanding of hedgehog populations?

Field projects

Inter-tidal gastropods

The inter-tidal gastropods *Hydrobia ulvae* and *Hydrobia ventrosa* are illustrated in Figure 7.6. *Hydrobia ulvae* is often said to occupy more frequently inundated, more brackish conditions than its close relative *H. ventrosa*.

Examine the reliability of this statement in an area of inter-tidal mud flat or salt marsh, using belt transects across tidal creeks or seawards down tidal flats (Check tide tables and obtain local information on the speed of the incoming tide before tackling this type of project.)

Temporal and spatial variations in molluscan populations

Use the Lincoln index to identify variations in molluscan frequencies within the study area according to seasonal and daily weather conditions; in protected, shaded, moist habitats and adjacent open, exposed habitats such as walls. Do the results correspond with the predictions suggested by reading Section 7.10?

Animal distributions and nutrient transfers

On moorlands or in large fields and paddocks, grazing animals such as sheep, cattle, horses, goats, or rabbits may be responsible for the transfer of significant quantities of nutrient from one area of the moor/paddock to another, with significant impacts on the pasture vegetation and soil structure. This happens because several species preferentially graze in one area but defaecate in another. Further complications are introduced as a result of some species preferring to eat as individuals, whereas others eat as groups.

Design and carry out a survey to compare the different patterns of movements of these species over a 24 hour period and compare these patterns with information obtained on the location and abundance of droppings. What are the implications of these patterns for the application of fertilizers in fields?

Try to detect any associated variations in plant species type or soil surface variations. Many aspects of this data set are appropriate to the types of statistical analyses described in Chapters 11–13.

Part Three

Mapping and Aerial Photography

8 Mapping

8.1 Introduction

The construction and use of accurate maps are essential tasks for most field scientists. Maps are essential if we are to understand the factors controlling the distribution of plants and animals; the impacts of human activities upon them; important ecological processes such as migration, succession and speciation; evaluation of the resource potential of poorly known areas; development of rational management plans for biological reserves and national parks; control of the spread of disease; or study of the effects of pollutants in the landscape.

In the following three chapters we concentrate on three inter-related and complimentary sets of survey and mapping methods. Chapter 8 – field mapping methods appropriate to small areas of less than a few square kilometres. Chapter 9 – survey, mapping and cartographic techniques appropriate to studies of individual species or particular communities at the local, regional, national or international scale. Chapter 10 – the use of oblique and aerial photography and satellite imagery for both small and large area studies. In practice the methods may be used on their own or in combination.

Data sources

The choice of data sources or mapping technique usually involves making a compromise between the following factors:

1 the objectives of the study;
2 the nature of the phenomena to be mapped;
3 the character, accessibility and extent of the pre-existing information;
4 the size and shape of the study region;
5 the availability of topographic maps;
6 the number and skill of the survey team;
7 the time and funds provided for field and laboratory study;
8 the provision, scale, character and quality of aerial photography and satellite imagery;
9 the accepted level of precision and reliability of the completed map.

It is possible to map many different aspects of the biology of an area: e.g. the presence or absence of particular taxa (varieties, species, families etc.); the relative frequency or actual abundance of these taxonomic units in an area; various aggregations of these species as plant and/or animal assemblages, communities, biogeocenoeses, or plant formations; the important characteristics of individuals or groupings such as the extent of tree crown damage or infestation, leaf shape and size, productivity, growth rate, date of flowering, mortality, rate of spread or retreat.

The potential use to which these methods may be put can be illustrated by an example from the vegetated dunes of the Coorong coast of South Australia (Gilbertson and Foale, 1977). Here, there occurs a spectacular purple flowered, coarse-leaved shrub called the kangaroo apple, (*Solanum laciniatum*). This shrub is a member of the nightshade family and is currently being investigated as a source of solasidine – a precursor of the very important cortico-steroid medical drugs (Alcock and Symon 1977). It is often hidden from aerial survey by the overstory of gums and wattles, and consequently has to be located by ground survey. Observations of the presence or absence of this species combined with related information on environmental, competitive and successional factors may help to determine those factors which most influence distribution and hence provide clues to those environments elsewhere in which the kangaroo apple might best be cultivated. Additional data on frequency, reproductive rate, age-mortality and associated species would assist in the identification of those environments in which optimum conditions occur. If cultivation were to prove too difficult or expensive, these data will also indicate the locations within the natural distribution, at which the species might be exploited most effectively, with minimum problems of access and cost.

Proxy data

The mapping of proxy or surrogate data is sometimes necessary when dealing with secretive or nocturnal animals. For example, if it is important to establish the present status of deer in an area of coniferous forest or plantations, then problems of time, accessibility and money, may cause the survey team to concentrate on the distribution and abundance of animal droppings, tracks or characteristic damage to the tree trunks. The problems of working in difficult terrain may also lead the workers to adopt measures of abundance other than frequency per square kilometre. Instead they may record frequency per standard search time in that type of terrain.

In under-developed parts of the world, surveys intended for wildlife management may not concern themselves (primarily, if at all) with the species composition of each habitat or terrain type. Instead the mapping studies may concentrate on the structural complexity and foliage characteristics of the vegetation (Section 4.5). These biological properties are important for economic development, since they influence the ease with which the area might be penetrated by vehicles and permit the construction of roads. Similarly for conservation planning, effort might be concentrated on establishing the number, size, shape and complexity of habitat types; factors which have a very important bearing on conservation potential and are more readily determined than species composition.

8.2   Field mapping of small areas

Degrees of precision

The degree of precision required in survey and mapping will be determined by the objectives of the survey and will depend upon the methods available as well as the practical problems encountered on the ground. Whereas mapping in a small 'urban' wildlife refuge might expect to locate the position of a plant with a degree of precision closer than 1 m in 100 m, and to estimate the altitude to within 0.01 m, the survey team in a remote forest may recognize less than 20 per cent of the plant species present and would be satisfied if they knew their position on the ground to within 1 km.

The need for very detailed maps of small areas arises for many reasons. For example, the available topographic maps may be out of date. Coastal salt marshes and sand dune systems are often important wildlife reserves. In the past they have experienced, and nowadays continue to experience, considerable changes in physiography as a result of both natural processes and human activity. There are many possible causes of change. People are continuously changing the shape, form and properties of land by clearing, planting, fertilizing, accelerating erosion and modifying depositional environments. Entirely new habitats are created at ground level in reservoirs, spoil heaps and gravel pits, whereas tall buildings and quarry faces generate new examples of the once rare cliff habitat.

Three types of survey procedure are especially useful for small area surveys. The grid survey is a development of the detailed 1 m × 1 m quadrat approach to vegetation studies. The offset survey will readily produce reliable maps of linear features such as hedges, walls, streams or strand lines. The most versatile methods employ a plane table. Plane table survey is capable of dealing with areas as small as 5 m × 5 m up to those determined by maximum visibility and the curvature of the earth's surface.

Distance measurement is easily achieved by four methods: pacing, tape, optical range finder and the more sophisticated tachaeometric facilities of a level.

The determination of height relationships is often crucial to understanding the field relationships of plants and animals. This may be accomplished at little cost with an Abney Level, or a home-made or purchased surveying level.

Preparations for field mapping

Most survey and mapping procedures are very simple. However, in the field, things can and do go wrong unless some very simple rules are followed:

1   List all the equipment likely to be required before you set out and check it is available or with you. Obtain identification keys, floras, maps or aerial photographs as necessary. If precipitation is likely, then use drawing film rather than paper for your field mapping. Drawing film will not tear, rip, shrink or dissolve in rain. It will retain the mark of a hard or soft pencil in all weather conditions.

2   On site, sketch the general distributions of all the important biological or topographic features. Determine the size and shape of the area to be mapped (a simple compass and pacing survey may be useful) and identify how the map is to be orientated at the chosen scale, in order to cover the study area in only one, or as few sheets as possible. If practicable, check that the features shown on previously published maps are incorporated into your study, to ensure your survey may be related to earlier surveys and/or the National Survey Grid.

3   With your colleagues, decide upon the appropriate notation and symbols for mapping and the procedures for communication in the field (people are often out of earshot).

4   Plan the pattern of moves of the survey lines or plane table(s) in advance to avoid the practical problems of mapping and negotiating creeks or streams, and maximize the area covered at each stage.

5   Ensure there are always at least two fixed points in common between two maps of adjacent areas; this will ensure that maps can be joined satisfactorily.

6   Always place the following data on the map sheet immediately before you start the survey: the direction of magnetic north, a line scale to indicate distance (avoid writing the scale on the map in the form 1:100 etc., subsequent reductions or enlargements will make nonsense of such a record), the date, the full details of the site with its name, location, National Grid reference or latitude and longitude, and the object of the survey. Details of ownership, access and management will be useful in the future.

7   Always map in the general outlines of the area and the species distribution first.

8   If the survey board is to be moved or traversed, fix the position of the new survey point on your plan before you move the survey point to that location.

9   Annotated, panoramic 'instant' photographs taken with a Kodak or Polaroid camera in the field before and after the survey will enable a close check to be made on any questions concerning the relative positions of the survey stations, traverses and biological data, after the survey and will help prevent problems resulting from misidentification of survey locations.

10   Never leave points to be joined up in the laboratory or classroom after leaving the field. Complete as much work as you can whilst the features are in view.

11 Check the accuracy of the survey by taking measurements between features at each stage in the construction of the map and noting if the distances are accurately portrayed at the scale of the map. Refer to fixed points where position has been independently determined (e.g. triangulation points, bench marks, established buildings). Determine the closing error (see below), and include a statement of accuracy on the map (e.g. accuracy is better than 1 m in 100 m). If worse than this, repeat the survey!

Time spent on maximizing the efficiency and quality of survey is never wasted. If it can go wrong it will go wrong. Make sure that a field mapper reading your map in twenty years time will know what has been mapped, why it was mapped, where and when it was mapped, and the reliability of the map.

Grid and offset survey

Grid and offset surveys are illustrated in Figures 8.1(a) and (b). Such surveys may in principle be carried out at any scale, but accuracy is obviously determined by the dimensions of the grid. The 1 m × 1 m quadrat is probably the smallest grid commonly used in vegetation studies. The spatial capacity of the grid and offset survey will be limited by the use of 30 m tapes which tend to restrict surveys to 30 m either side of the survey and base points. The length of the survey base line may be in excess of several hundreds of metres, depending on the precision with which a bearing can be followed and accuracy checked.

Equipment

30 m tapes; plastic knitting needles or pegs; string for grid cells; compass; cross staff (see Figure 8.2), alternatively either a compass, a right angle prism, or string arcs for determining a right angle; marker posts (bamboo) for offset baselines; graph paper with clipboard or large notebook with gridded pages; marker pen for labelling rows, columns and offsets.

Procedure

The survey procedures are illustrated in Figures 8.1(a) and (b). It is essential to obtain true right angles for grid and offset surveys. A simple cross staff or the use of 3:4:5 and 9:12:15 triangles will define a right angle as illustrated in Figure 8.2.

1 Once the outline is established, progressively draw in the remainder of the map.
2 Survey points of particular interest should be numbered, while individual species may be labelled with a capital letter, and communities or other features occupying larger areas can be given various shading types and symbols.

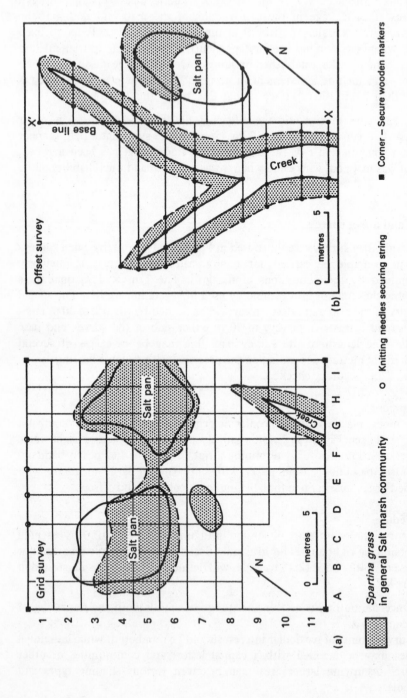

Figure 8.1 Grid and offset surveys of the pattern of invasion by *Spartina* grass of a coastal salt marsh in southern England (Note: if the creek in Figure 8.1(b) had been slightly broader, the offset baseline X–X¹ could not have been laid out.)

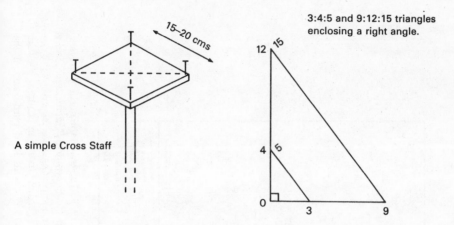

Figure 8.2   Devices for setting out a right angle

3 Always move forwards towards the next area to be mapped. This avoids the problem of having to study and map trampled areas. Wet land is particularly sensitive to trampling and must be mapped with great care.

Relative merits

Figure 8.1 suggests how the grid survey method has a much greater capacity for detailed mapping when compared with the offset method. However, the presence of surfaces which cannot be pinned (e.g. rock, creeks, lakes) may render the technique impracticable. In general, grid surveys are well-suited to complete novices. There are no judgements required on which measurements need to be taken and thus little is missed through surveying error.

Offset surveys are ideal for rapid, less detailed situations or the study of linear features such as hedges, streams and footpaths. Over long distances inaccuracies are likely to creep in, for example the presence of ridges may make it very difficult to lay out a straight offset baseline by eye. The location of both ends of the base line should be fixed independently from reliably established survey points. Care must be taken to ensure gaps do not occur in the survey as a result of missing opportunities to lay out offsets.

Case study: Changing vegetation zonation around an English lake
Differences in tolerance and competitive ability between various aquatic and semi-aquatic plant species at North Fen at the head of Esthwaite Water in the Lake District have led to a marked vegetation zonation (Figure 8.3). Sedimentation at the mouth of Black Beck has induced rapid successional changes, leading to a marked displacement of the vegetation zones in 1929 compared with 1914–15.

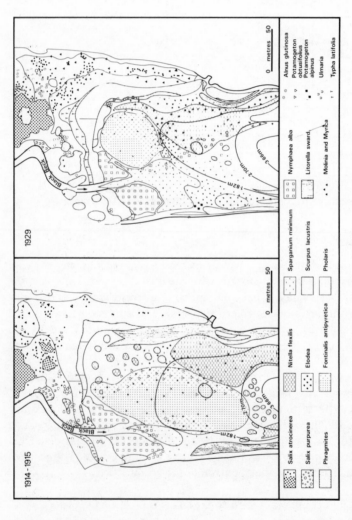

Figure 8.3 Vegetation zonation at North Fen, Esthwaite Water, English Lake District; 1914–15 and 1929. The fen to the north of the lake is divisible into three zones: (a) a strip bordering the Black Beck with rapid sedimentation; with *Calamagrostis, Phalaris* and locally *Filipendula, Salix purpurea* and *Salix decipens;* (b) a central strip of moderate sedimentation and carr comprising *Salix atrocinerea;* (c) an eastern strip of very low sedimentation with *Molinea caerulea* and *Myrica gale.* In 1929 note the effect of sedimentation in changing the lake bed topography and vegetation zoning. (Redrawn from Pearsall (1918) and Pearsall in Tansley (1939); with kind permission of the Cambridge University Press.)

The survey methods used by Pearsall (1918) and Tansley (1939) were essentially offset surveys. The north–south orientated shore was used as a baseline. A marker rope was placed along the offset, and the vegetation and depth recorded along the transect at right angles to the shore baseline. The deeper water vegetation was recorded after sampling it with a grab auger.

The scope and scale of this survey are suited to offset or plane table survey. The study has emphasized the rapidity of ecological change that can be brought about in some lakes by deposition from sediment-rich streams.

8.3 Plane table surveys

Plane table survey is a simple, fast and effective survey method and is illustrated in Figure 8.4 and Plate 8.1.

Equipment
Plane table and tripod; alidade; rule (may be incorporated into the alidade); spirit level; ranging posts (bamboo painted alternately in red or white for each 0·5 m); 30 m tapes; drafting tape; best quality drawing paper (or drawing film for use in wet weather).

The plane table is essentially a drawing board covered with drawing paper or film. Plane table boards typically measure 500 mm square or 600 mm × 400 mm. They are expensive to purchase new but are readily manufactured from 23 or 30 mm blockboard. The surface may be improved by using a filler. The board will not warp. Ideally the board will swivel and tilt on a ball and socket joint mounted on a tripod and then be held in position with a locking screw. Extendable tripod legs are also used to obtain a perfectly horizontal surface. Whilst these are all useful features, ultimately, all that is required is a level board or table.

The busiest working part of the equipment is the alidade. This is a sighting device, along which a straight line can be drawn (see Figure 8.4 and Plate 8.1). A typical manufactured version has a slot in a collapsible plate at one end, through which the surveyor looks to see a wire sight, in a slot, at the other end. Home-made alidades using a rule are perfectly satisfactory for simple work as long as the rule does not warp and the pins stay upright. The line of sight through the pins must be parallel with the edge of the rule. Errors will occur if right and left-handed people alternate in the use of the alidade on a plan, because they will differ in their selection of which side of the rule is chosen to draw the sighting line or 'ray' on the map.

Distance determinations may be made by tape, optical range finder or the tacheometric distance measuring facility built into many levels (see below). In some studies, tapes are unsatisfactory. Errors may be introduced by sloping ground, tape sag, or handling problems, especially in bad weather. A cheap and most useful alternative is the optical range finder. The most familiar

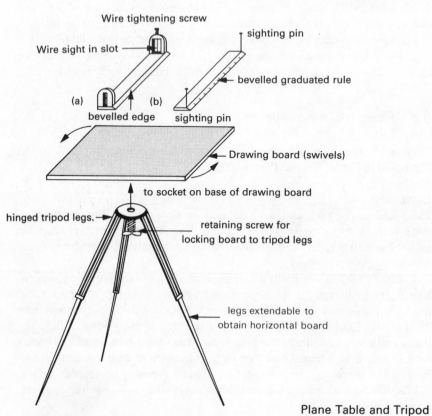

Alidades
(a) 'standard' model
(b) 'home made'

Wire tightening screw

sighting pin

Wire sight in slot

bevelled graduated rule

(a)   (b)
bevelled edge   sighting pin

Drawing board (swivels)

to socket on base of drawing board

hinged tripod legs.

retaining screw for
locking board to tripod legs

legs extendable to
obtain horizontal board

Plane Table and Tripod

Figure 8.4 Alidades, plane table and tripod for a plane table survey

employ the same principle as the split image focussing systems of some 35 mm cameras. Two images are brought together and focussed, the distance is then read off the calibration scale. When used over longer distances the optical range finder has important advantages of speed and ease of use, and as a result fewer errors.

Procedure
There are four common mapping procedures used with plane tables:

1 radiation;
2 traversing;

Plate 8.1 Plane table, alidade and automatic level in use during a vegetation survey in the Yorkshire Dales. The hand calculator is being used to determine distances based on the tacheometric distance measurements from the level. (Photo: M. Kent)

3 intersection;
4 resection.

The last is a method for precisely relocating positions on the ground determined by previous surveys. It is useful but beyond the scope of this text.

Radiation and traverse
The procedures are illustrated in Figure 8.5.

1 Determine a centrally located survey point (e.g. site I, Figure 8.5), with good, uninterrupted views of the features to be mapped.
2 Secure the paper or film to the drawing board by wrapping the paper over the board and securing it with drafting tape, not drawing pins or sellotape.
3 Use a spirit level to set the table horizontal, ensuring the board is appropriately orientated in terms of the shape of the study area. Mark on the direction of magnetic north with an arrow, the date, a line scale and the other relevant information.

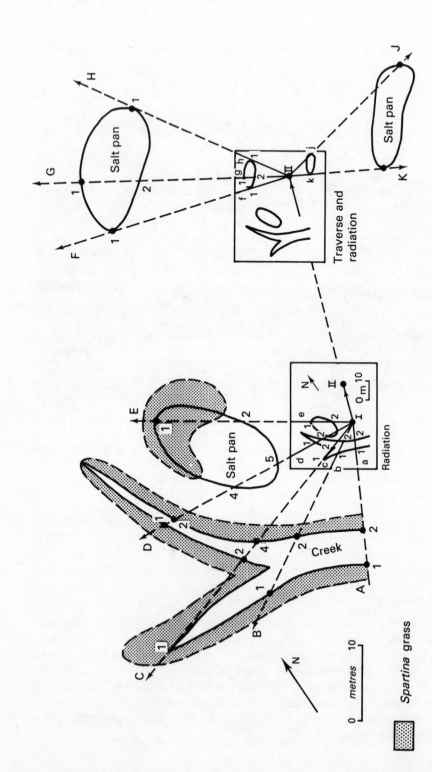

Figure 8.5   Plane table survey by radiation and traverse

4   Mark site I on the plane table map so that its position corresponds with its approximate location within the study area (Figure 8.5).
5   Place a sharp point on site I on the plane table and slide the edge of the alidade against the pencil. Line up the alidade sights with the bamboo ranging rod marking the target. Draw a faint ray or line from I along the edge of the alidade towards the target.
6   Measure the distance from the centre of the plane table to the target with a tape or range finder, and mark off this distance along the drawn ray on the map. The position of the target is then fixed relative to site I. Continue and fill in between the mapped points by freehand interpretation on the map.
7   If a target cannot be readily mapped from the first location, a traverse to a second location may help.
8   Select a second main survey station on the ground, labelled site II on the map shown as in Figure 8.5. Fix its position with the alidade and tape or range finder.
9   Move the board to site II. Set it up again, ensuring that it is orientated correctly by sighting back to site I from site II with the alidade. A second check on the accuracy of the mapping of the second survey position is to use the compass and measure the new direction of magnetic north and check if it agrees with the former direction. Plot in the targets as before.
10   Finish by traversing back to the first survey point – site I. Any locational error which has crept in during traverses will be determined as a closing error. In the study, the position of site I as first mapped, and its position as mapped at the end of the survey, may be found to differ on your plan. This difference is the closing error of the survey. If possible the map should then be re-plotted with the error distributed according to the length of the traverses made. (This is only possible if the traverses and main survey points are mapped before detail is added to the board.) Note the accuracy of the survey on the plan.

Intersection
The principle of intersection is illustrated in Figure 8.6. Sites, plants or features are located by rays from either end of the measured baseline I–II. The method is very accurate, rapid, and has the great advantage that inaccessible features can be mapped from afar.

The procedures are essentially similar to those used in radiation (Figure 8.5), except that the only distance measured is that of the length of the baseline. However, care is needed when setting up the board to ensure that it is correctly orientated.

1   Use the arrow previously drawn towards magnetic north to re-align the plane table at each end of the survey baseline.
2   Correctly and fully label the rays to the targets from the two ends of the baseline. Otherwise, upon moving the plane table and mapping at the other end of the baseline, confusion will occur – there being little indication of which ray pointed to which feature.

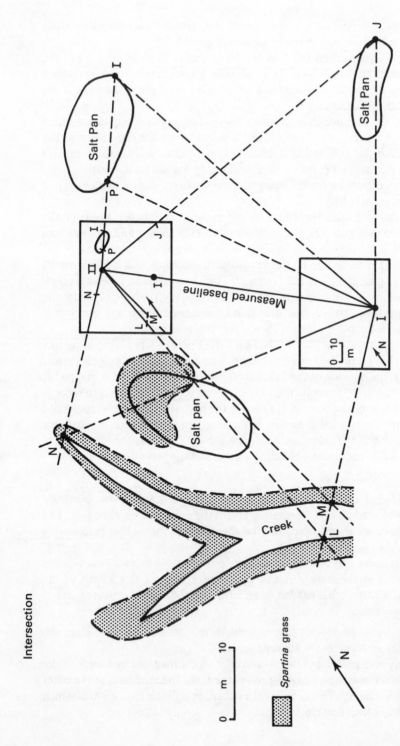

Figure 8.6 Plane table survey by intersection

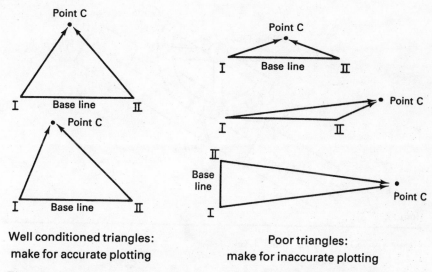

Well conditioned triangles:
make for accurate plotting

Poor triangles:
make for inaccurate plotting

Figure 8.7 Well and poorly conditioned triangles forming triangulation nets

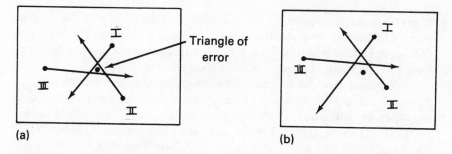

(a)

(b)

Figure 8.8 Triangles of error, (a) true position lies within the triangle of error formed by rays sighting on to one point; survey points I, II, III; (b) true position lies outside the triangle of error

3   Ensure that the triangulation net is well conditioned as shown in Figure 8.7. Rays intersecting at angles between 45 degrees and 135 degrees make for greatest accuracy.
4   Check the closing error and re-distribute if possible. If a third base line is used the rays converging on one feature will define a triangle of error as shown in Figure 8.8. This will indicate the accuracy of the survey.

Contouring
A detailed contour map may be produced at the same time as the plane table survey by using a levelling staff as the target and determining height with a level or an Abney level mounted adjacent to the plane table. A few institutions may possess alidades which have levels or theodolites built into them.

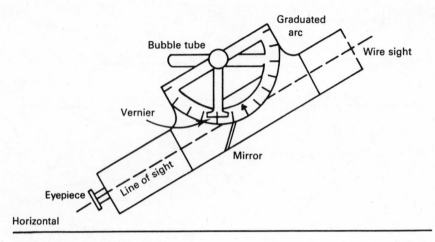

Figure 8.9　Sketch of an Abney level

8.4　Height determination

Every field scientist must be able to determine the height of one feature relative to another or to a datum point. The datum for Great Britain is the Ordnance Datum which is a mark representing mean sea level at Newlyn, Cornwall. Such datum points should be completely stable.

Altitude is a most useful environmental measure in many habitats. For example, the coastal zonation of salt marshes reflects many factors e.g. frequency of inundation, exposure, water salinity etc. Differences of a few centimetres may be critical for the survival of plant species (Dalby 1970). The relative magnitudes of the factors and their impact at a location are conveniently 'summarized' in a statement of absolute or relative altitude. Height determinations in biological or geographical fieldwork frequently are made with the use of an Abney level or a level and staff.

The Abney level
The Abney level illustrated in Figure 8.9 is used to determine the deviation of a line of sight above or below the horizontal. The Abney incorporates an index arm which is attached to a bubble with a vernier pointer. The vernier scale moves along a graduated arc which is fixed to the side of the sighting tube. The instrument may be held or placed on a supporting mount.

Altitudes are determined as shown in Figure 8.10. The angle of elevation or depression of the target is determined from the vernier scale and the distance to the target determined with a tape or range finder. The calculation of height is then a matter of simple trigonometry (Figure 8.10) using trigonometrical tables or a pocket calculator with scientific functions (Figure

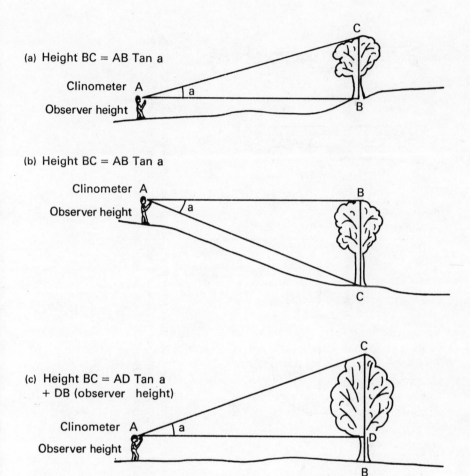

(a) Height BC = AB Tan a

Clinometer A

Observer height

(b) Height BC = AB Tan a

Clinometer A

Observer height

(c) Height BC = AD Tan a
+ DB (observer height)

Clinometer A

Observer height

Figure 8.10 Determination of height with an Abney level. Angle a is read from the Abney level's graduated arc

8.10). Abney levels or clinometers are not difficult to manufacture with a spirit level and protractor.

8.5 Principles of levelling

Levelling is an accurate and effective method of height determination. It may be carried out with a level and staff (Figure 8.11 and Plate 8.1) or less reliably with an Abney level (Figure 8.9), or a home-made levelling tube made from a cardboard tube or rod attached to a spirit level.

The principle is illustrated in Figure 8.12 and an example of the results is

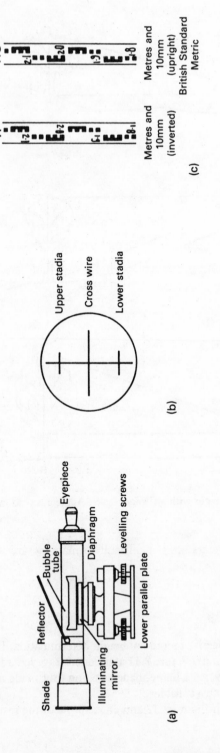

(a)

(b)

(c)

Figure 8.11 (a) Sketch section through a Dumpy level; (b) Cross wires and upper and lower stadia, as seen through the level eyepiece; (c) Metric reading staves. Divisions are in units of 0.01 m. Alternate metres are in black and red. Some levels do not automatically correct inverted images produced by the level in the process of magnification. In this case the inverted staff **R** is useful.

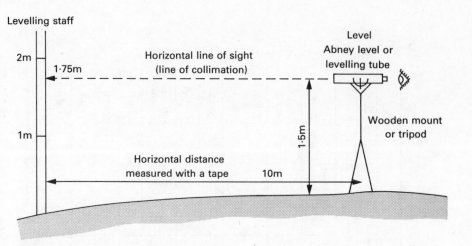

Figure 8.12 Determination of relative height by reference to a horizontal line of sight
Since the line of sight is horizontal and 1.5 m above ground at the observer's position,
the ground surface must have fallen by 0.25 m below the base of the tripod by the time
the staff is reached 10 m distant.

given in Figure 5.6. The target is a levelling staff, which is a wooden or
aluminium staff up to 3 or 4 m long, and graduated in metres and 10 mm units
(Figure 8.11(c)). It is viewed through the eyepiece of the level which is marked
with a crosswire. The level is set up so that the line of sight represented by the
cross wire is horizontal. Relative elevations are then determined by identifying
their height above or below that horizontal line of sight. This line is termed the
line of collimation or collimation axis (Figure 8.12). These height data are
recorded on standard log sheets (Table 8.1).

It is not necessary to know the actual height of the horizontal line of sight at
the position of the observer to solve a height determination problem. Usually
the field worker determines the difference in height between two or more
points. If one of these points is a position of known altitude, such as an
Ordnance Survey bench mark or trigonometrical point, then the height of each
point above or below a nationally fixed datum can be easily calculated.

For example, it is clear that if the height of the line of sight at the position of
the surveyor is 1.5 m above the ground surface, then a reading for the
horizontal line of sight of 1.75 m on the staff, must indicate that the ground at
the staff has fallen 0.25 m compared with its height at the position of the level.
Should the height of the starting point A be unknown, it is usual to allocate an
arbitrary height datum e.g. 100 m for the purposes of calculation. This
arbitrary datum should be sufficiently high to avoid encountering negative
figures in the ensuing calculations. When the height is related the to Ordnance
Datum, this is referred to as a reduced level and the height is quoted in metres
OD i.e. above Ordnance Datum at Newlyn, Cornwall.

Table 8.1 *The rise and fall entry method for recording levelling data*

Site	Survey instrument		Date
Grid ref. (initial)			Surveyor
(final)			Closing Errors

Backsight	Intermediate Backsight	Foresight	Intermediate Foresight	Ground Rise	Ground Fall	Reduced level	Upper Stadia	Lower Stadia	Distance	Notes
2.60										Bench mark on wall
						20.00				20.00 m O.D. Grid ref. Point A
			1.50	1.10		21.10				Point B
		1.00		0.5		21.60				Point C
										Abney moved to position 2
2.10						21.60				Point C
	1.6			0.50		22.10				Point D
		2.4			0.8	21.30				Point E
0.8										Level moved to position 3

Notes: Use the backsight and foresight data to calculate rise or fall in ground level; then add this rise or fall to the previous sites reduced level to obtain the reduced level of the new site.

A 'sighting-back' towards a levelling staff is termed a backsight, as illustrated in Figure 8.13. Sighting forward is termed a foresight. Intervening locations are termed intermediate sights.

Beginners will find that it is easier to interpret their results if they record their survey diagrammatically as shown in Table 8.1. With increasing familiarity the data will be intelligible when entered on to a log sheet (Table 8.1) or a surveyor's log book. It is wise to draw up the profile or plot the contours onto a map at the earliest opportunity.

Distance measurement

Distances may be determined by tape, optical range finder or the tacheometric distance measuring incorporated within the optical systems of the level. The latter system is very easy to use. The height difference between the upper and lower stadia seen through the level's eyepiece (Figure 8.11(b)) as read on the staff is multiplied by 100. This gives the distance in metres from the staff to the level.

Procedures and checks

To obtain reliable results observers must ensure that:

1 the staff is upright – check by using the spirit level on the staff;
2 the line of sight is horizontal – check the spirit level is level, errors occur when people lean on, or bump into the level's tripod. A further check is to ensure that the height differences from the line of sight to the upper and lower stadia are the same;
3 when moving the staff, ensure that the level or Abney does not move;
4 the traverses are completed with out-and-back loops. This will permit the closing error to be calculated, which in turn will indicate the accuracy of the survey. The closing error is defined as the distance between the true or adopted height at the starting point, and the height of that point as calculated at the end of the survey traverse. If the closing error is within acceptable tolerance limits then it should be distributed in proportion to the lengths of the traverse distances involved, and the heights of B, C, D etc., adjusted accordingly.

8.6 Levels

Three types of levels are likely to be encountered: dumpy levels, quickset levels and automatic levels. Beginners will find they can understand the principles of levelling by making a simple levelling tube from a cardboard roll. Levels are essentially telescopes which are mounted on tripods (Figure 8.11 and Plate 8.1).

In all cases the device must be set up to give a horizontal line of sight seen through the crosswire. The old-fashioned dumpy level has three or four levelling screws which are adjusted to ensure the line of sight is horizontal in all

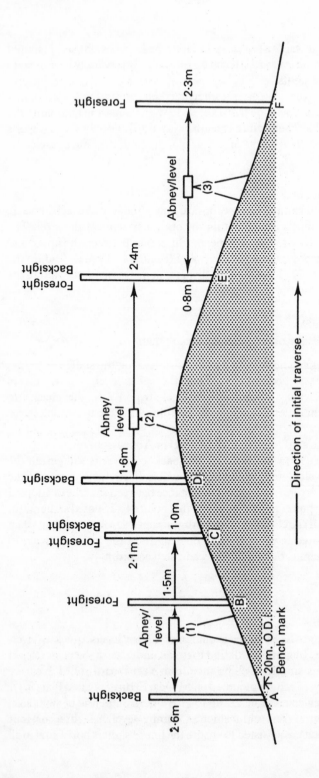

Figure 8.13   Levelling along a transect. The line of sight through the level at position (1) is 22.6 OD. The ground surface at C is only 1.0 m below the line of sight; therefore the ground surface at C is 21.6 m OD. Distances along the transect may be measured by tape or tacheometry.

directions, the built in spirit level serves as a guide. The quickset level is set up more rapidly than the dumpy. The telescope is set horizontal with the aid of the ball and socket and built-in spirit bubble. Fine adjustments in the direction of the line of sight are made using a tilting screw. Automatic levels are set up using the ball and socket joint and spirit bubble, and/or levelling screw. A built-in compensator ensures that the line of sight is horizontal when the bubble is in the centre of the bubble dome.

8.7 Projects

Classroom projects

Changes in vegetation zonation
Pearsall's maps of the vegetation and lake bed topography of North Fen in the English Lake District are shown in Figure 8.3.

1 Using maps and/or profiles constructed on transparent overlays or tracing paper, identify the areas of greatest topographic change on the lake bed in the period between 1914 and 1929.
2 Are sedimentation rates highest in the vicinity of Black Beck?
3 Compare this rate with your observations of the average sedimentation rate around the margins of this lake.
4 Discuss in detail, the vegetational response to this pattern of sedimentation.
5 To what extent would any future management plan for the lake have to take into account or even attempt to influence activities outside the lake shores? Justify your views.

Field mapping project

1 Determine the relationships between plant and animal zonation and elevation around lakes, salt pans or coastal salt marshes. Elevation integrates the effects of changing frequencies of water-logging, exposure, inundation, elimination of light etc.
2 Investigate the pattern of invasion of coastal salt marshes by *Spartina anglica*. This species has been extensively transplanted about the globe (Figure 1.3). It has been observed that on the west coast of Britain, *Spartina anglica* appears initially to colonize areas of salt marsh not occupied by other plant species (e.g. salt pan, creeks and channels, and the most seaward margins of the marsh). Once pioneer clones have developed in these areas, the species then appears to invade the marsh by seeding. The remarkable capacity of this species to spread and then trap silt may lead to a rapid transformation of the marsh. Other marshes may be colonized as a result of propagation from drifted fragments of leaf or stem. Devise and carry out a mapping programme to test these ideas.

9 Distribution mapping of large areas

9.1 Introduction

There is a natural curiosity to know what happens to the distribution of plants and animals outside the area which is well known to you. When it is necessary to establish the distribution of a particular taxon or community over hundreds or thousands of square kilometres, special procedures must be adopted. For example, many people are likely to become involved in the project and there may be problems of language or difficulties in communication. In addition, at these spatial scales the character of the ground mapping unit, within which individual surveys are being conducted, becomes extremely important.

In the earliest surveys of this type, the mapping units used nearly always reflected political-social-administrative units. This is often the case today, but there has been a world-wide trend towards adopting grid mapping techniques which are more useful to the biologist and geographer. Within Great Britain and Ireland, this is illustrated by the gradual replacement of the Vice-County mapping system by Grid mapping surveys with increasing levels of spatial resolution.

9.2 Vice-County mapping

The counties of the British Isles, like the states or provinces of other parts of the world, vary widely in size, shape, land use and physiographic diversity. These factors complicate comparisons of the biological resources recorded in the various County Floras, Faunas or archives. Consequently, as the sophistication of biological and geographical questions about these resources has increased, so there has been a trend to develop smaller, more manageable and more comparable mapping units. The grid mapping techniques applied in Chapter 3 to Rutland and Essex are the development of methods devised in the mid nineteenth century. The grid surveys replace the Vice-County system which was proposed for Great Britain by H. C. Watson (1847–1859 and

1873–4) and extended into Ireland by C. C. Babbington and R. L. Praeger (Praeger 1906). The most recent account of this mapping system is given by Dandy (1969).

The Vice-County system represented a half-way stage between the abandonment of biological mapping using administrative units such as those defined by county boundaries and its replacement by more systematic and ecologically useful units. Despite more recent developments, many biological survey archives are still held and illustrated in Vice-County form. The locations of the Vice-Counties are shown in Figure 9.1 and their names are listed in Table 9.1.

There were several reasons for the abandonment of the Vice-County system. Problems arose concerning the artificial distortion of the distribution of the organisms. One example is shown in Figure 9.2. The map shows the distribution of the common and distinctive land snail *Acanthinula aculeata*, which is found in leaf litter in woods and hedges. Comparison of the data on presence/absence when plotted according to the Vice-County system with those recorded on 10 km × 10 km grid squares (Figure 9.3), shows how the former exaggerates the distribution, especially in Scotland, Ireland, Lincolnshire and the West Midlands.

The problem arises first because no information is presented on the abundance of *Acanthinula*, the maps being constructed for presence/absence data only. Second, the variations in the size and shape of the field mapping units represented by the Vice-Counties cause the larger Vice-Counties (e.g. 63, South-West Yorkshire) or Vice-Counties with more diverse topography, climate, geology, soils and land use etc. to have a much higher probability of containing a particular species than a correspondingly smaller and less diverse Vice-County.

These early Vice-County surveys nevertheless have undoubted importance for ecological research. In particular, they enable comparisons to be made between contemporary and past distributions, although for several reasons, care must be exercised in their interpretation, especially as it concerns likely agents of change. Problems can be caused by inexact knowledge of the Vice-County boundaries. Early workers often experienced difficulty in defining or recognizing the boundaries in the field. Second, there was always the temptation to slightly expand the mapping unit in order to incorporate a new record within the Vice-County surveyed by that natural historian, rather than allocate it to an adjacent area. Third, rivers often formed Vice-County boundaries and it was not always clear to which side of the river channel a new find should be allocated.

Fourth, unfortunately many of the early surveys and resultant maps also represented the cumulative fund of knowledge about an area. Thus, local extinctions were obscured, as were changes in species' habitat preferences and distributions. New species were always added, while those previously recorded, but not found in subsequent surveys (possibly having become locally extinct), were too often assumed to have been missed through chance or survey error.

Table 9.1 *Vice-Counties of Great Britain and Ireland* (From Dandy (1969); the Channel Islands may also be numbered 'C'; while the Isles of Scilly may be regarded as a separate unit 'S'.)

Vice-Counties and Vice-County numbers, England and Wales

1	West Cornwall (with Isles of Scilly)	26	West Suffolk	52	Anglesey
2	East Cornwall	27	East Norfolk	53	South Lincolnshire
3	South Devon	28	West Norfolk	54	North Lincolnshire
4	North Devon	29	Cambridgeshire	55	Leicestershire (including Rutland)
5	South Somerset	30	Bedfordshire		
6	North Somerset	31	Huntingdonshire	56	Nottinghamshire
7	North Wiltshire	32	Northamptonshire	57	Derbyshire
8	South Wiltshire	33	East Gloucestershire	58	Cheshire
9	Dorset	34	West Gloucestershire	59	South Lancashire
10	Isle of Wight	35	Monmouthshire	60	West Lancashire
11	South Hampshire	36	Herefordshire	61	South-East Yorkshire
12	North Hampshire	37	Worcestershire	62	North-East Yorkshire
13	West Sussex	38	Warwickshire	63	South-West Yorkshire
14	East Sussex	39	Staffordshire	64	Mid-West Yorkshire
15	East Kent	40	Shropshire (Salop)	65	North-West Yorkshire
16	West Kent	41	Glamorgan		
17	Surrey	42	Breconshire	66	Durham
18	South Essex	43	Radnoshire	67	South Northumberland
19	North Essex	44	Carmarthenshire		
20	Hertfordshire	45	Pembrokeshire	68	North Northumberland (Cheviot)
21	Middlesex	46	Cardiganshire		
22	Berkshire	47	Montgomeryshire	69	Westmorland (with North Lancashire)
23	Oxford	48	Merionethshire		
24	Buckinghamshire	49	Caernarvonshire	70	Cumberland
25	East Suffolk	50	Denbighshire	71	Isle of Man
		51	Flintshire		

Vice-Counties and Vice-County numbers, Scotland

72	Dumfrieshire	86	Stirlingshire	100	Clyde Isles (Buteshire)
73	Kirkudbrightshire	87	West Perthshire (with Clackmannanshire)	101	Kintyre
74	Wigtownshire			102	South Ebudes
75	Ayrshire	88	Mid-Perthshire	103	Mid Ebudes
76	Renfrewshire	89	East Perthshire	104	North Ebudes
77	Lanarkshire	90	Angus (Forfar)	105	West Ross
78	Peebleshire	91	Kincardineshire	106	East Ross
79	Selkirkshire	92	South Aberdeenshire	107	East Sutherland
80	Roxburghshire	93	North Aberdeenshire	108	West Sutherland
81	Berwickshire	94	Banffshire	109	Caithness
82	East Lothian (Haddington)	95	Morayshire (Elgin)	110	Outer Hebrides
		96	East-Inverness-shire (with Nairn) (Easterness)	111	Orkney Islands
83	Midlothian (Edinburgh)			112	Shetland Islands (Zetland)
84	West Lothian (Linlithgow)	97	West-Inverness-shire (Westerness)	113	Channel Isles
85	Fifeshire (with Kinross-shire)	98	Main Argyllshire		
		99	Dunbartonshire		

Table 9.1 *contd.*

Vice-Counties and Vice-County numbers in Ireland

H1	South Kerry	H15	South-East Galway	H29	Leitrim
H2	North Kerry	H16	West Galway	H30	Cavan
H3	West Cork	H17	North-East Galway	H31	Louth
H4	Mid Cork	H18	Offaly (King's	H32	Monaghan
H5	East Cork		County)	H33	Fermanagh
H6	Waterford	H19	Kildare	H34	East Donegal
H7	South Tipperary	H20	Wicklow	H35	West Donegal
H8	Limerick	H21	Dublin	H36	Tyrone
H9	Clare	H22	Meath	H37	Armagh
H10	North Tipperary	H23	West Meath	H38	Down
H11	Kilkenny	H24	Longford	H39	Antrim
H12	Wexford	H25	Roscommon	H40	Londonderry
H13	Carlow	H26	East Mayo		
H14	Leix (Queen's	H27	West Mayo		
	County)	H28	Sligo		

9.3 The 10 km x 10 km grid

Current biological mapping of taxa in the British Isles is organized by the Biological Records Centre of the Institute of Terrestrial Ecology. This body is part of the Natural Environment Research Council and is based at the Monk's Wood Experimental Station, Abbots Ripton, Huntingdon. They employ the 10 km × 10 km grid first employed in Britain by the Botanical Society of the British Isles in 1954. An example of the resulting map is shown as Figure 1.1 for *Spartina anglica*. The map is based on the Ordnance Survey National Grid which is printed on the 1:10,000, 1:25,000, 1:50,000 and 1:63,360 sheets.

Field workers are issued with standardized field survey cards and checklists for each 10 km × 10 km grid square. The presence of a species in the square is recorded with a dot. The maps for the higher plants were published by Perring and Walters (1962) as *The Atlas of the British Flora*. A set of transparent 'environmental overlays' are also presented to accompany the maps. The atlas represents a milestone in biological survey. A re-survey has been underway for some years and aims to establish the precise character of the distributional changes which have occurred since 1960.

A second major publication using the 100 square kilometre grid is the *Atlas of the Non-Marine Mollusca of the British Isles* edited by M. P. Kerney (1976). This atlas gives dot maps for 199 species of mollusc living in Great Britain and Ireland. Typical maps are shown in Figures 9.3 and 9.4. Transparent environmental overlays can be obtained from the Institute of Terrestrial Ecology (see Appendix).

These surveys are important data sources. However the diversity of the British landscape in a 10 km × 10 km grid square may cause a masking or

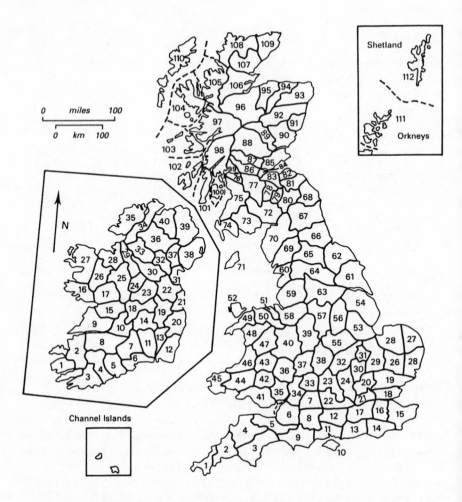

Figure 9.1   Vice-County boundaries in Great Britain and Ireland

averaging of notable biological variation within one square. Consequently there have been a number of attempts to use much finer grids as illustrated by the Rutland study in Chapter 3.

9.4   The UTM grid

At the other end of the scale there has been pressure to understand the distribution of plants and animals at a continental scale. A number of committees of European biologists have adopted the International Universal Transverse Mercator (UTM) grid with its 50 km × 50 km grid square as the base unit for the biological mapping of Europe. UTM grids are now published on maps of many countries published at the scales of 1:1,000,000 or 1:500,000.

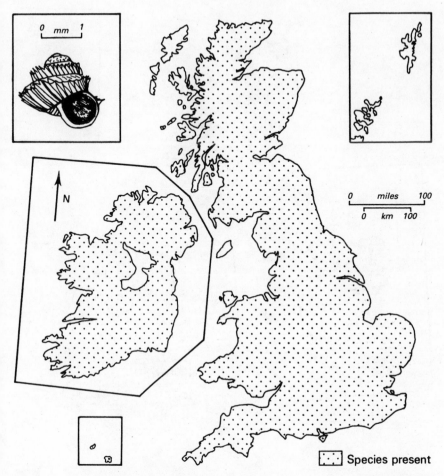

Figure 9.2 The distribution of the land snail *Acanthinula aculeata* from all records in Kerney (1976), and replotted according to presence or absence in the Vice-Counties of Great Britain and Ireland

Presence or absence of species is recorded by competent biological surveyors for each grid square.

The value of objective and systematic surveys of large areas, such as the continent of Europe, is illustrated in Figures 9.4 and 9.5. These show the British distribution of the readily identified land snail (*Pomatias elegans*) plotted on the 10 km × 10 km grid and a provisional model of its distribution on the UTM grid. In Britain this species is mainly restricted to lime-rich soils and regions with infrequent frosts. However, when the distribution is examined on a European scale, two further interesting points emerge. First the presence of the species in the very different acidic soils and climate of the Massif Central of France is obviously anomalous, and warrants further

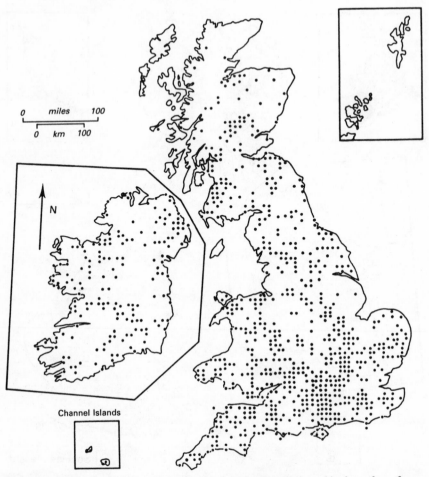

Figure 9.3   The distribution of all records of the land snail *Acanthinula aculeata* from Kerney (1976), plotted on the 10 km × 10 km grid system (Redrawn by kind permission of Dr M. P. Kerney and the Biological Records Centre, Institute of Terrestrial Ecology.)

investigation. Second, there is seen to be a population in eastern Denmark, well to the north of its main distribution: the question arises, is this a relic population or an advance guard?

World distributions

The identification and explanation of the distributions of individual species, genera or families on a global scale is fraught with problems. There are difficulties involving the scale of the finished distribution map and the very generalized environmental information with which they can be compared (see Figure 1.3). Geographers, in particular, might feel that this was the scale at

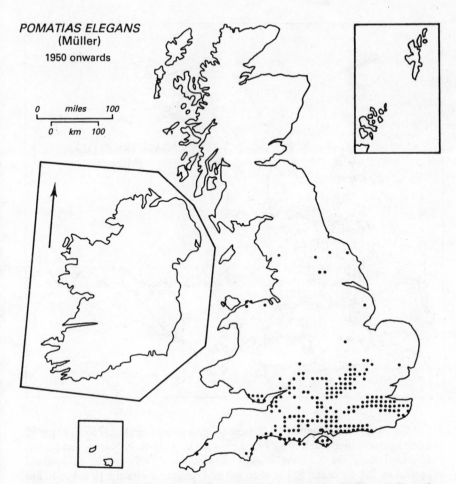

Figure 9.4 Present British distribution of the frost sensitive land snail *Pomatias elegans* plotted on the 10 km × 10 km grid square system (Redrawn by kind permission of Dr M. P. Kerney and the Biological Records Centre, Institute of Terrestrial Ecology.)

which they could make their most important contributions. However, at global or even continental scale, interpretation may be hindered by a lack of reliable information on identification or taxonomy. Reliable interpretation needs a full knowledge of the taxonomic relationships between the species and the limitations of the species concept, especially where the species studied are widely separated. For example, the common reed (*Phragmites communis*) has a mild tolerance of brackish water, which combined with its ability to disperse into and colonize fresh and brackish water habitats, has led to this species being widely distributed around the globe. Nevertheless caution and taxonomic knowledge must temper the desire to interpret. Although *Phragmites* is relatively uniform in appearance and behaviour throughout its

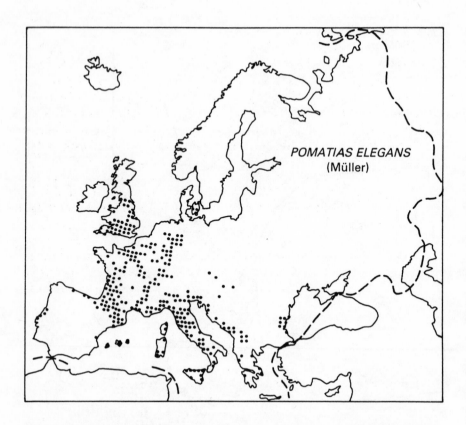

Figure 9.5   Present distribution of *Pomatias elegans* plotted on the UTM 50 km × 50 km grid square system (Redrawn by kind permission of Dr M. P. Kerney and Collins publishers.) Note the apparently wider distribution of this species in England suggested by the use of the 50 km × 50 km grid compared with the 10 km × 10 km grid in Figure 9.4.

range, this 'single' species may be found in the future to be made up of a complex of closely related forms, each adapted to special conditions (see Section 1.2 and Ranwell 1972). Factors involving genetics, evolution and physiology, as well as obvious climatic constraints, may be unseen but are important.

Analysis at global scale will also emphasize other controls on distributions:

1   natural barriers to migration such as seas, mountain ranges and ice caps;
2   the rapid rate at which plants and animals have been transported, accidentally and deliberately, by people about the world;
3   the significance of geological and climatic controls and evolution on biological distributions through the impact of plate tectonics, climatic change and sea level fluctuations.

9.5 Transplant studies

Transplant experiments offer a valuable means of studying environmental controls on a taxon at the world scale. One of the world's most widely travelled species is *Spartina anglica*, described in Section 1.2. The present distribution of this grass is shown in Figure 1.3. This distributional data led Ranwell (1972) to observe that the northern limit of the grass appears to be controlled by frost frequency. Field surveys showed that frost killed 99 per cent of transplants in Holland, while those in the milder south of Holland were virtually unaffected. However, the northern limit of *Spartina* in the southern hemisphere is at 35 degrees south, and appears to be controlled by day length. The plant appears unable to flower in the relatively short day conditions experienced during the spring, summer and autumn above these latitudes. Such transplant experiments are elegant tests of these types of hypotheses. The significance of such factors as annual air temperature or similar potential climatic controls, may tend to be over-emphasized as a result of the general lack of other, biologically important, environmental information at these macro- or meso-scales.

9.6 Projects

National and international distribution maps

Data sources
1 Kerney, M. P., *Atlas of the Non-Marine Mollusca of the British Isles* (NERC/Conchological Society of Great Britain and Ireland, Hills Road, Cambridge, UK, 1976).
2 Institute of Terrestrial Ecology, Environmental overlays, (NERC, 1976).
3 Perring F. and Walters M., *Atlas of the British Flora* (London, Nelson, 1962 and supplements).
4 University or College atlas.
5 Climatological atlas.
6 Soil Survey of England and Wales. *Soil map of England and Wales 1:1,000,000* (1975), Harpenden, Hertfordshire UK.

Aims and methods
(a) In Britain, it has been said that the land snail *Pomatias elegans* is mainly confined to lime-rich soils and is limited in its northern distribution by an intolerance of frost (see Figure 9.4). To what extent does this 10 km × 10 km grid map of its distribution support this contention?

1 Using the UTM survey, and climatological and pedological data from your atlas investigate whether or not frost continues to limit its distribution on the European mainland.

2   Identify the areas where it occurs as isolated pockets beyond the main distribution. Are these regions in any sense anomalous? Do they represent favoured habitats in which the snail might be favoured by some form of factor compensation? How would you establish whether these outlying areas are regions into which the snail is invading; are they residuals of a former more widespread, continuous distribution, possibly of a former period of milder climate?

(b) The woody climber *Clematis vitalba* (Old Man's Beard) is usually associated with neutral or alkaline soils. Assess the validity of this suggestion for England as a whole. Which soil types are most favoured, if any? Assess whether these observations are also justified at the 2 km × 2 km tetrad scale employed in the Rutland survey (see Figure 3.3e). Try to identify other environmental factors which may be important in this example.

(c) Describe and explain in climatic and/or pedological terms the distribution of the following plant species:

Pinus sylvestris   (the Scot's pine)
Tilia cordata   (the small-leaved lime)
Ulex gallii   (the western furze or gorse)
Ruscus ruscus   (the butcher's broom)
and the land snails:
Pupilla muscorum
Acanthinula aculeata
and the freshwater mollusc *Anisus vorticulus*.

(d) Isolated regions are often much less thoroughly searched than more accessible or populous areas. Inspect the numbers of molluscan species per 10 km × 10 km grid square in England and Wales as recorded in Kerney (1976) and investigate whether species numbers are fewer in more remote areas. Would you consider other human or natural factors likely to influence this pattern, if so which factors and why?

The interpretation of Vice-County distributions

Data sources
1   Roebuck and Boycott, *Census of British Non-Marine Mollusca* (Conchological Society of Great Britain and Ireland, 1921).
2   Ellis, *Census of British Non-Marine Mollusca* (Conchological Society of Great Britain and Ireland, 1951).
3   Kerney, *Atlas of the Non-Marine Mollusca of the British Isles* (NERC and the Concological Society of Great Britain and Ireland, 1976).
4   Perring and Walters, *Atlas of the British Flora* (Thomas Nelson and Sons, 1962).

Method
1 Search the atlases by Kerney and by Perring and Walters for examples like *Acanthinula aculeata* discussed previously, in which Vice-County mapping distorted or exaggerated a species' distribution.
2 Identify the changing distributions of the introduced molluscan species *Potamopyrgus jenkinsi*; *Marstoniopsis scholtzi*; *Menetus dilatatus* and *Hygromia cinctella*. To what extent might passage along rivers and canal systems by the aquatic species account for the present inter-drainage basin distribution of the freshwater taxa *Potamopyrgus*, *Marstoniopsis* and *Menetus*?

Transplant studies

The following taxa have been transported around the world either accidentally or deliberately: the common periwinkle, the hedgehog, many species of rat, the pigeon, house sparrow, starling, domestic cat and dog, rabbits, foxes, hares, coniferous tree species, marram grass (*Ammophila*) and Canadian pondweed (*Elodea canadensis*). An additional data source is Elton (1958).

Method
1 Using standard atlases and consultation with a local expert (see Chapter 3) try to identify the pattern of invasion of one or more of these species.
2 Attempt to determine which natural, cultural and economic factors have influenced the rate and direction of spread.
3 If you suspect that the taxon is now in a semi-stable equilibrium with environmental factors, try to identify those factors from an examination of your atlas.

10 Aerial photography and satellite imagery

10.1 Introduction

Aerial photographs and satellite imagery are important weapons in the armoury of the modern biologist and geographer. Four types of film are used: panchromatic black and white; infra-red black and white; colour yielding approximately 'true' colours; and colour infra-red (CIR), one type of which is often called 'false colour'. All of these are readily available from photographic stores or can be ordered from photographic suppliers.

Infra-red films need special care. They may be stored for short periods in a refrigerator at or below 13 degrees C (55 degrees F). Storage for periods in excess of two days should be in a freezer at temperatures below 0 degrees C (−18 degrees F). Two to four hours are needed for the films to warm up before use. They must be loaded and unloaded in absolute darkness.

The reasons for using these different types of film are easily grasped with an understanding of the electromagnetic radiation spectrum and the impact of plant surfaces on that radiation.

10.2 The radiation spectrum

Visible light is one part of a vast range of electromagnetic radiation emitted from objects in space (Figure 10.1). White light can be split into its component wavelengths by a prism in the classroom. A similar thing occurs in the atmosphere as a result of refraction of light by water molecules to give a rainbow. The human observer sees these different wavelengths of radiation as different colours; the shorter visible wavelengths being seen as ultra-violet, the longer visible wavelengths being seen as reds. Wavelengths which are shorter or longer are not visible to the eye. They may however be 'visible' to photographic films. For example, most commonly available photographic films are also sensitive to radiation in the ultra-violet. Others are sensitive to radiation in the near infra-red (Figure 10.1).

The advantage of infra-red sensitive films lies in their capacity to discriminate between vegetation types and levels of photosynthetic activity

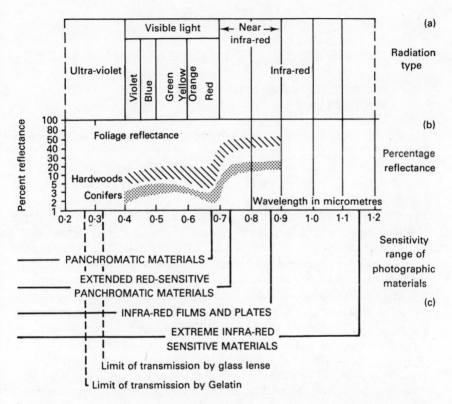

Figure 10.1 Inter-relationships between radiation type, percentage reflectance of broadleaf hardwoods and needleleaf conifers, and the sensitivity range of photographic materials (Redrawn with kind permission from Lyons and Avery (1977) and Kodak (1977).)

much more effectively than can be accomplished by eye or normal photographic films. Figure 10.1 shows that in the infra-red wavelengths the quantity and 'quality' of the light or radiation that is reflected back from plants (and hence be seen or photographed) depends upon the particular abilities of that plant or species to absorb, transmit and reflect the incident radiation. The reflectance of near infra-red radiation from leaf surfaces is between two and eight times greater than occurs in the visible part of the spectrum. These differences in spectral reflectance result from differences between the chlorophyll pigmentation of leaves, leaf surface effects such as leaf angle (to the sun), leaf shape, and the orientation of tissues within the leaf. It is not surprising that considerable differences exist between the near infra-red reflectance characteristics of hardwood and conifer leaves. Consequently, when photographed from afar, they are much more easily distinguished on the prints of infra-red films than normal panchromatic types of film. The reflectivity of leaves can also alter according to the physiological status of the plant, so that colour infra-red photographs are often used to detect diseased or dying trees of a certain species.

10.3   Panchromatic (black and white) and true colour photography

The advantage of colour film derives from the fact that the eye can distinguish between 200 and 500 colours and tones, whereas it can only discern between 10 and 40 grey tones on black and white prints. (See back cover. The top photograph is a black and white print of satellite imagery in the infra-red waveband 800–100 nm of South Australia; compare with colour infra-red imagery below. Bottom photograph is colour infra-red imagery of South Australia from the Landsat satellite (Source: Landsat I observation 15001–0001; Image centred 34° 40′S, 139° 14′E; 20 December 1972: Air Photographs Pty Ltd, 620–624, Burwood Road, Auburn, Victoria, Australia 3124).)

Despite its more restricted resolution, black and white aerial photography is widely used because it is very cheap and requires less skill in camera exposure setting. In many parts of the world, black and white aerial photography taken in World War II, represents the only detailed data source available.

A typical high angle, oblique, black and white photograph is shown in Plate 10.1. This is interpreted in Figure 10.2(a). There is clearly a marked plant zonation within and around the lake. Such zonations are widespread (see Sections 2.6 and 5.6). The actual pattern and width of the vegetation belts depend on the shape of the original basin, the steepness of its sides, the pattern and rate of sedimentation, and the impact of past and present human activity. A detailed planimetric map of the area is given in Figure 10.2(b). This may be combined with either the aerial photograph, or the interpretation, to make a true scale plan of the vegetation belts using the 'slide projector method' (see below).

Carr is a north country word for wet woodland. It has been adopted widely by ecologists to describe this type of vegetation. The fen carr (e) is distinguishable from the abundant purple willow carr (f) by its darker tone. It is tempting to map in further purple willow carr immediately west of the overflow of Cunsey Beck as it leaves the lake. The zone of sedges or fen grasses (d) is distinguishable only with difficulty from the reed swamp of *Phragmites* (c). This interpretation requires verification by ground survey. This is true of all surveys which use aerial photography or satellite imagery as their main data base. Such aerial photographs tend to emphasize the distribution of plant communities, whereas the ground studies by Pearsall of North Fen immediately to the north of Out-Dubs Tarn (Figure 8.3) record the distribution of individual species or specimens.

The advantage of colour photographs compared to panchromatic photographs is that they often discriminate more clearly between types of vegetation, soils, rock outcrops, soil moisture status, patterns of burning, salinization and similar phenomena.

Plate 10.1 Oblique aerial photograph of plant zonation at Out-Dubs Tarn near Esthwaite Water, English Lake District An explanation is given in Figure 10.2 (Photograph reproduced with the kind permission of Professor J. K. S. St Joseph, Cambridge University Collection, Copyright reserved.)

10.4 Infra-red photography

Black and white infra-red panchromatic films often give enhanced subject contrast and object brightness, as a result of the lesser scattering by air of radiation from the reflecting surface to the camera, when compared with photographs taken only in visible wavebands. This in turn produces greater haze penetration on one hand, and adds a certain dramatic quality to the photographs on the other. Lighter tones on such photographs result from the greater reflectance of near infra-red radiation from the surface compared to the degree of reflectance of radiation in the visible part of the spectrum. Broad-leaved deciduous tree leaves reflect more near infra-red radiation than the absorbent needle leaves of conifers. Similarly in aquatic ecosystems, silt-laden water, marsh, swamp or shorelines all stand out as light tones against the dark or black tones of the near infra-red absorbent water.

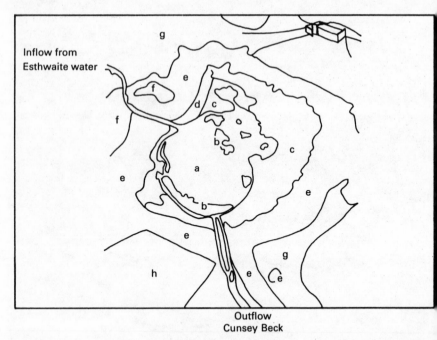

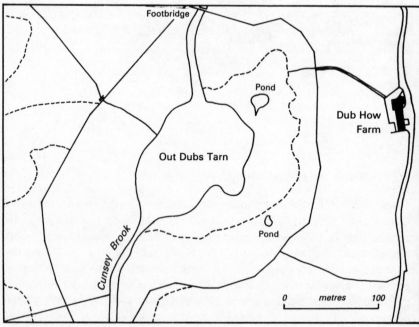

Figure 10.2   (a) Interpretation by Godwin of the plant zonation of Out Dubs Tarn near Esthwaite Water, English Lake District, shown in Plate 10.1 (Reproduced by kind permission of Sir Harry Godwin.) (b) Ordnance Survey plan of Out Dubs Tarn (Redrawn and reproduced by kind permission, Ordnance Survey, Crown Copyright reserved.)

The types of results obtained by colour infra-red films are illustrated on the back cover. This 'false colour' photography has been derived from imagery obtained by the sensors of the Landsat satellite. As long as simple rules concerning exposure setting and focusing are followed, excellent CIR photographs may be taken by anyone from the ground, a vantage point or an aircraft, using a standard 35 mm camera with an orange or red filter (a Wratten 25 or 29 filter). Wratten filters are readily available from Kodak dealers.

This type of film is best known for its ability to detect camouflage and dead or diseased vegetation. Healthy, unstressed, photosynthesizing plants actively absorb and use radiation which is reflected by dead, unhealthy or 'imitation' plants. Consequently it is possible in many situations to compare the metabolic state of plants and penetrate camouflage. On CIR film the healthy, actively photosynthesizing plants have leaves which show up in splendid shades of magenta or red. Stressed plants become paler on CIR photography as they reflect more radiation back towards the camera, whereas diseased or dead plants show as blue.

An example of infra-red satellite imagery

The photos on the back cover were taken by Landsat 1 of part of South Australia. The former is a black and white print of radiation reflected in the 8–1100 nm – the near infra-red band (nanometers = 0.8 to 1.1 micrometres = 0.8 to 1.1 millionths of a metre).

The second photo on the back cover is a colour equivalent of the adjacent area of reflected radiation. The advantages on CIR are clear from the comparison. Many features stand out with greater clarity on CIR. The irrigated pasture lining the lower River Murray appears as a sinuous, bright red strip, approaching the black (energy absorbing) Lake Alexandrina to the north of the Coorong Lagoon, adjacent to the Southern Ocean. The higher rainfall areas of the Adelaide Hills and the Mount Lofty Ranges are seen as a broad, red band running NNE–SSW. Black areas are reservoirs or burnt ground. The stressed vegetation and the urban areas in metropolitan Adelaide show up as pale blue, whereas the well-watered parks and playing fields of the city form a bright red square around the Central Business District. Well-tended suburban gardens and parks show in a dull red. The vast range of stress conditions associated with water shortage and heat experienced in the paddocks (fields) to the north of the city and beyond the irrigated lands of the Murray are shown as areas of blue, white and pink speckles. Possible tree disease in the Adelaide Hills and die-back in the coastal swamp, show up as blue patches and lines.

The equivalent black and white prints of the radiation recorded at shorter wavebands (500 to 600 nm – green light; and 600 to 700 nm – red light), do not show the above vegetational details with such clarity.

10.5   Cameras

Ground or aerial photography may be attempted with almost any camera. 35 mm cameras with see-through-lens (SLR Single Lens Reflex) metering systems are reliable and easy to use in the field. If near infra-red photography is intended, an orange or red filter is necessary to prevent radiation of shorter wavelengths entering the camera, an additional hand-held light meter will be useful but is not essential; the technical specification of the hand-held meter is often better than SLR meters.

Correct exposure setting is essential for all colour and colour infra-red photography (see below). When photographing from light aircraft, shutter speeds should be faster than 1/300th second and preferably up to 1/500th second to avoid vibration problems. Shutter speeds in excess of 1/1,000th second are not required because colour films may not perform satisfactorily at these very fast speeds.

Most good camera lenses will serve for near infra-red photography. However, unless the camera lens has been corrected (achromatized), there will be a difference between the visual focus position appropriate to one's eyes, and the infra-red focus position appropriate to the near infra-red film and the orange or red filters. The appropriate focus position for near infra-red is usually shown on 35 mm SLR cameras by a red mark on the distance focus position scale of the camera. If this mark is absent, the appropriate focus position can be identified by trial and error photographs taken with one roll of film.

This trial and error approach will also resolve any initial problems over exposure values and shutter speed. The main principle is to take the photographs at what is believed to be the correct exposure setting for that shutter speed, and then rephotograph one half and then one whole stop on each side of the original reading. Note the various combinations and assess the final results in that light.

Transparencies

Colour prints have their uses, but transparencies (slides) in 'natural' colour or near infra-red colour have three distinct advantages:

1   there is the possibility of easily changing magnification during projection to clarify site detail;
2   plans can be readily made by sketching the projected detail onto a sheet of paper affixed to the projection wall or board;
3   they facilitate the direct use of the 'slide projector method' (see below) for making a map directly from oblique photographs.

10.6 Taking oblique aerial photographs

Panchromatic and colour photography

Film; obtain as fast a film as possible – at least ASA 50, preferably faster, i.e. a higher ASA number.
Filters; a strong haze filter will help.

Procedure

1 If in light aircraft, secure all maps, films and other equipment; determine the best view, noting the effects of light and shadow.
2 Focus the camera on infinity.
3 Secure the lens focus position with masking tape, if necessary.
4 For black and white film use a 2× yellow (minus blue) filter; for colour use a strong haze filter.
5 If in a light plane set shutter speed as fast as possible, e.g. 1/500th second.
6 Determine the appropriate exposure setting with SLR metering or hand-held meter.
7 If in a light plane, then stop down one 'f' stop from the light meter reading and take photograph.
8 Locate the photographed area on a map, noting direction of the shot.

When in a light aircraft, try to take the photograph as near to the vertical as possible! This results in easier plotting of the photographed features, gives better definition and avoids horizon or sky light upsetting the exposure reading for the ground shot.

10.7 Colour infra-red photography (CIR)

Film: Kodak film IEI 35-20, Kodak Ektachrome Infra-red Film – this film yields transparencies from which prints may be taken. This film must be processed by Kodak.
Filter: Wratten no. 15 (orange).

Procedure for ground based and oblique photography

1 For large area views set focus to 15 to 27 m, or use the red infra-red focusing indicator if available on your camera.
2 Set the film speed to ASA 100 on the camera control and determine the correct exposure; hand-held light meters are best, however SLR meters are usually perfectly adequate.
3 Cover the lens with an orange filter.
4 Take the photograph.
5 Rephotograph, one half 'f' stop up and one half 'f' stop down.

6   To be totally certain, rephotograph again one 'f' stop up and down.

In the early stages, trial and error may be needed to determine the most appropriate exposure settings for particular cameras and filters. However, at least one of the three photographs taken in the initial bracketing procedure should be satisfactory. It always helps to try to use exposures with lens openings of f/11 and smaller, for sharp images and improved depth of field.

10.8   Map production from oblique photographs/ transparencies: the slide projector method

Useful maps may be produced using oblique photographs taken from the air or high vantage points, using the 'slide projector' method of Harris and Haney (1973), when:

1   the landscape is 'fairly' level;
2   good base maps are available, with several photographed features actually located on the maps;
3   the photograph angle is not too oblique; the nearer the vertical the better. Usable maps can be obtained with photographic angles as low as 25 degrees to the horizontal.

Equipment
Good base maps; transparencies (slides) of the oblique photographs; slide projector; tracing paper, a board.

Alternatively make transparent overlays and use an overhead projector: the 'keystone' property of these machines can be turned to good advantage.

Procedure
1   Make a tracing of those details found on the topographic base map which also appear on the transparency. Mount the tracing on to a board.
2   Project the oblique aerial photograph transparency on to the board, using as long a focus position as possible. Alter the angle of the board until the common boundaries on the transparency and traced map coincide. Fix these positions and transfer the biogeographical information on to the traced map.

Inevitably there will be some problems of focus and blurring of lines. The courageous use of the pencil is required. More fastidious students will find the projective net method of map construction from aerial photographs described by Howard (1970) to be inherently more satisfactory, if a little more complex.

10.9   Mapping very large areas

In many parts of the world aerial photographs or Landsat imagery represent both the document on to which biological and geographical data are mapped and also form major data sources in their own right. Large area surveys are often carried out as part of a Natural Resources Inventory Establishment programme. While study of the vegetation may form a most important part of that programme, the photographs will also be used to provide information on soils, landforms, recent ecological changes, rivers, settlements and land use.

Vertical aerial photographs and photomosaics

The clarity with which vegetation can be mapped from black and white vertical photographs may be appreciated by inspecting Plate 10.2, which shows the vegetation around the Waiowa Volcano, Papua New Guinea, some eleven years after its 1943 eruption. The vegetation mapping of large and remote areas such as this is achieved by the compilation of a photomosaic of vertical aerial photographs of the same scale.

These mosaics are made up of the central, less distorted, areas of photographs. Mapping units are then identified on the basis of similarities of texture, tone, and patterns of certain sizes and shapes on the photographs. These units are then visited by ground survey teams who identify the vegetation types responsible for the patterns seen on the photographic image. In the case of Papua New Guinea such ground truth data were established for some 40 per cent of the region mapped. This took over 20 years (Paijmans, 1975)!

10.10 Landsat imagery

An example of a vegetation and land utilization map produced from Landsat imagery is given in Figure 10.3(a). These results may be compared with the original 1885 Land Surveyor's maps of the area which are redrawn in Figure 10.3(b). The most recent vegetation map of the area produced from conventional sources by Carnahan (1976) is reproduced here as Figure 4.5(a).

Landsat 1 was launched in 1972, Landsat 2 in 1975, Landsat 3 in 1978 and Landsat 4 in 1982. Numbers 1 and 2 no longer function. Information is now also available from the French SPOT and Japanese Earth Reconaissance satellites.

The Landsat satellites operate in circular, nearly polar orbits at an altitude of 930 km. Each orbit takes 103 minutes. The plane of the orbit advances slowly around the Earth so that the 14 orbits completed in one day are equally spaced around the globe. The 15th orbit on the following day lies adjacent to the 1st orbit of the first of the previous day. This systematic shift in the orbit plane results in the same area of the Earth's-surface being covered every 18 days. The Landsat 1 satellite was not operational over all areas of the globe.

The Landsat 1 satellite, which yielded the data shown on the back cover, carried two sensor systems. The first was a four channel multi-spectral scanner (MSS) which recorded reflected radiation in the spectral regions which are

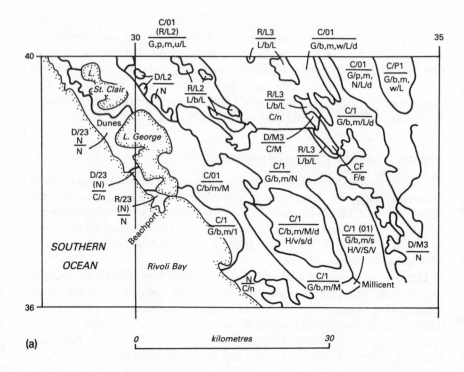

Figure 10.3(a) Vegetation and land use map of the Beachport region, South Australia, produced from Landsat imagery: the key is presented in Table 10.1 (Redrawn with permission from the Penola sheet, accompanying Heyligers *et al.* (1978); CSIRO Technical Memorandum 78/1; Division of Land Use Research.)

illustrated in Figure 10.1; 500 nm–600 nm which is primarily green light; 600 nm–700 nm (red light); 700 nm–800 nm (red – near infra-red); and 800 nm–1100 nm (near infra-red). These are known as bands 4, 5, 6 and 7 respectively. The second sensor system comprised three television camera systems with each camera recording reflected radiation in one of three wavebands: 475 nm–575 nm (green light); 580 nm–680 nm (red light) and 690 nm–830 nm (red–near infra-red). This second system was modified and then discontinued on the later satellites.

A swath 185 km wide is imaged by the MSS in the four wavebands as the Landsat satellite progresses along its orbital path. The instantaneous field of view of the scanner is square and results in a picture which is made up of individual ground cells each of which is 79 m × 79 m across. This basic cell is called a picture cell – hence it is commonly referred to as a pixel.

The reflected radiation sensed for each 79 m × 79 m ground cell is converted into digital form as a series of numbers in the range 0–63. These digital data are sent back to earth as radiowaves where they are received and processed at a ground receiving station. Here they are recorded and stored on magnetic tape.

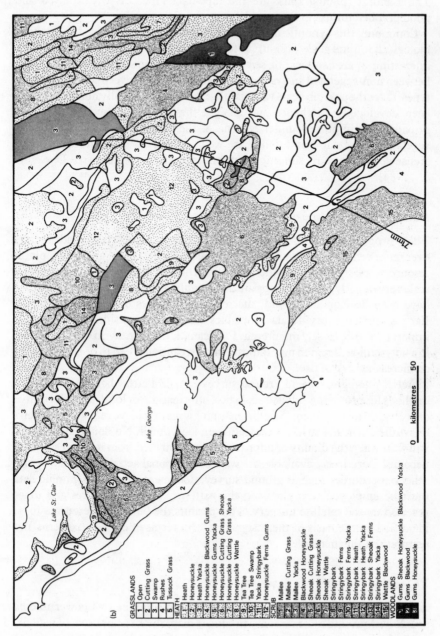

Figure 10.3(b) Vegetation of the Beachport area, South Australia, derived from the Land Valuation Survey (1885).

The Landsat 'photographs' are the tape-data reprocessed, displayed and photographed for use in field and laboratory.

Commonly this synoptic imagery is combined with established methods of biological, soil and geological survey for further study. The results are especially interesting in areas of arid or semi-arid terrain where strong contrasts occur between non-vegetated and vegetated areas, as well as between some vegetation types. Over the past ten years various other techniques, usually numerical, have been developed to enhance and analyse the Landsat imagery. These are providing an improved understanding of the character and distribution of vegetation, habitat conditions, as well as water, soil, geological and archaeological resources in important but inaccessible regions (see Dorsett *et al.*, 1984).

Advantages of satellite imagery

These properties result in Landsat imagery having two particular advantages over more conventional data sources for mapping biological and geographical resources. First, it can provide useful information at regional or continental scale very quickly, once appropriate data handling and interpretive procedures have been developed. Second, the repeated passages over each part of the Earth's surface every 18 days, enables features of considerable economic importance to be closely monitored. For example, an increase in dry grass cover in a hot, semi-arid region may indicate that precautions need to be taken against an increasing risk of fire.

Aerial photography and ground survey remain better than satellite based methods for providing detailed ecological information for local natural resource or ecological management. Satellite mapping aims to achieve regional overviews or syntheses, rather than concentrating on local detail. No doubt this balance will shift when the quality of information now obtained from military satellites becomes more freely available. As with conventional aerial photography and other data sources such as ground survey, documentary or cartographic data must be employed to understand the patterns, colours, textures and tones revealed on the satellite imagery. This permits the biological resources to be identified on the basis of their 'signature', (in terms of colours, patterns and textures) on the imagery.

Vegetation and land use descriptors

A major problem has been to determine how to combine this powerful data source with ground information to produce a map which can be used for regional survey by field workers, planning authorities, or for the day to day management of important problems such as fire or drought. Several aspects of this problem have been tackled in an important and pioneering study in South Australia which developed a methodology which is easily adapted to other locations.

At this regional scale, it is necessary to make many compromises in order to produce a map of value to regional planning agencies, including conservation bodies. Whereas the details of local soils, vegetation or land use can be usefully interpreted and studied for a small reserve from satellite imagery, this quantity of detail when presented for an entire region, would swamp the analyst with so much essential, and non-essential information that it would be almost impossible to plot the data in an interpretable manner at the map scales necessary to cover a very large region. Consequently the information from the imagery must be summarized and plotted on to mapping units with a basic grid unit much coarser than the 79 m × 79 m pixel of the original imagery. In South Australia the decision was made to adopt a 1 km × 1 km grid square as the basic ground mapping unit. At this scale it is possible to map from the imagery:

1 the principal land use types – their size, shape and distribution of the areas involved which are of great importance for biological conservation and management (see Chapter 15);

2 the major and minor natural vegetation types in each 1 km square grid cell.

The details of the categories employed are shown in Table 10.1. These data are combined to form mapping units. These are defined using a 'fractional notation' which is illustrated in Figure 10.4. The major and minor vegetation types are written with code letters, termed 'descriptors', as the upper part of the fraction (i.e. the numerator of the fractional notation). These upper two sets of descriptors record:

1 the significance of past and present human impact on the vegetation (known as its cultural status) which is of great importance for future management and conservation;

2 the structural and species composition of the vegetation, using code letters from the Australian structural formation vegetation description system (see Section 4.6).

For example, U D/L Z indicates that both low open forest and open heath occur in the area, the former is now effectively undisturbed or fully recovered from past disturbance; the latter, although disturbed, still retains its original structure and composition.

Land use data form the denominator or lower half of the fractional notation. Land use is described under the following headings:

1 land utilization; its products or the purposes of the land holding;
2 the average size of holdings;
3 any special features or properties of economic and ecological significance.

For example, F3e/V indicates that the principal land use comprises land holdings in excess of 2000 hectares, growing exotic softwoods for forestry purposes.

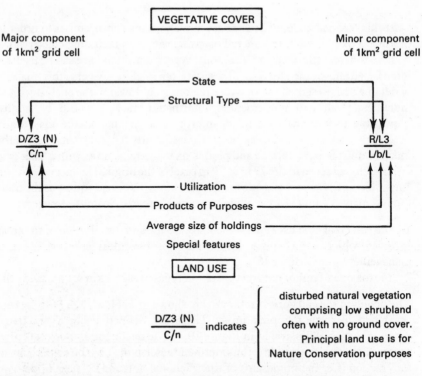

Figure 10.4   Fractional notation for combining vegetation and land use data into a mapping key for Landsat imagery (Reproduced with kind permission of CSIRO, Canberra, Australia.)

Procedure

1   Obtain satellite imagery (see Appendix for addresses).

2   Trace areas of distinctive and similar colour/tone/texture on to an overlay or tracing paper.

3   Calibrate these traced units by ground survey, or documentary or cartographic evidence. In the South Australian example the following main data sources may be used.

Vegetation	*Land Use*
Carnahan, (1976)	Heyligers *et al.*, (1978)
Landsat imagery	Laut (1978)
Laut *et al.*, (1977)	Laut *et al.*, (1977)
Specht, (1972)	South Australian Department of Lands, (1973)
	South Australian State Planning Authority, (1970, a, b; 1974; 1975, 1977)
	Twidale *et al.*, (1976).

The method of using fractional notation combining vegetation and land use descriptors is applicable to studies being carried out in any continent, (except of course in completely unvegetated areas with no land use, such as those in the middle of a large ice mass). The actual descriptors in use will of course have to be altered to meet the particular circumstances of the region under study.

10.11 Projects

Classroom projects

Mapping from oblique aerial photographs
1 Use the data on plant zonation about Out-Dubs Tarn, the Ordnance Survey plan (Plate 10.1, Figure 10.2(a)/(b)) and the slide projector method, to produce a true scale map of vegetation for the area.
2 Where do the greatest errors in location occur? Try to account for them.

Interpreting aerial photographs – Waiowa Volcano, Papua New Guinea
1 Identify the watersheds on the photograph, (Plate 10.2).

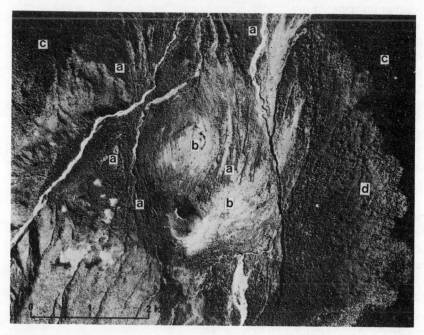

Plate 10.2 Vertical black and white aerial photograph of Waiowa Volcano, Papua New Guinea, eleven years after its eruption in 1943. The main pioneer species on the slopes of the volcano are *Casuarina* **a** and grasses **b** mainly *Saccharum spontaneum* and *Imperata cylindrica*. On the alluvial plain, right and top left, the extent of the devastation caused by the eruption is marked by the abrupt change from large crowned forest **c**, to *Octomeles*-dominated pioneer forest **d**. (Reproduced with kind permission of CSIRO, Australia, and the Surveyor General, National Mapping Bureau, Papua New Guinea.)

2   Can you identify the direction in which most devastation was caused?
3   What factors might have influenced this pattern?
4   From which direction does recolonization appear to be proceeding most rapidly?
5   In what manner might the possible answers to 3 and 4 be linked?

Satellite mapping and the changing pattern of vegetation and land use, South Australia, 1885–1978
Figure 10.3(b) shows the vegetation of the Beachport area of the South Australian coast as mapped during the original Government land surveys. Figure 10.3(a) reveals the present principal types of vegetation and land use as recorded on Landsat imagery.

1   Using the land use mapping key given in Table 10.1, evaluate the cultural status of the area's vegetation.
2   Identify which vegetation types are in the 'most natural' state. Try to account for their survival.
3   Try to work out why there might be correlations between the original distribution of vegetation types and (a) patterns of contemporary land use; (b) the size of contemporary land holdings. Use the two maps to establish the extent to which correlations may be substantiated.

Field projects

Infra-red photography
1   In your local area, investigate whether black and white infra-red film will distinguish between hardwoods and conifers in different lighting conditions (e.g. bright sunlight, partly clouded sky, overcast sky, low angle sun).
2   Is the ease of distinction affected by the age of the trees; the particular species involved; distance from the camera?
3   Trees, shrubs and bushes are often markedly wind-pruned in exposed coastal locations. Hypotheses might be: (a) this is due to increased mechanical abrasion/breakage on the upwind side, and/or (b) higher levels of photosynthetic activity on the sheltered downwind side.
(a) Investigate by establishing the ratio of sound to damaged twigs on exposed and protected sides of several trees/shrubs.
(b) Investigate by using CIR photography.
(c) Repeat for marram grass (*Ammophila arenaria*) on coastal sand dunes.
4   Atmospheric pollution. Identify the extent to which unprotected trees and shrubs at sites variously upwind or downwind, from a pollutant source (or particular sides of them) reflect stress. Using CIR photography and several photographers around the trees, take simultaneous photographs of the trees and evaluate the methodology.

Table 10.1 Vegetation and land use descriptors for use with Landsat imagery of South Australia (After Laut et al. 1977.)

Vegetative cover		Land use			
State	Structural type	Utilization	Products or Purpose	Average size of holdings	Special features
N virtually without cover e.g. Saltpan	M3 open forest M2 woodland L3 low open forest L2 low woodland L1 low open woodland	B rotation farming G grazing L light grazing	a barley h wheat t oats y rye b beef d dairy products m lamb/mutton w wool p pork	D 20 ha S 20–99 ha M 100–499 ha L 500–2499 ha V 2500 ha	d drainage i irrigation
U undisturbed or fully recovered vegetation – composition and structure intact	S3 open scrub S2 tall shrubland S1 tall open shrubland Z3 open heath Z2 low shrubland				
D disturbed natural	C3 chenopodioid scrub C2 chenopodioid shrubland G3 grassland G2 open grassland	H horticulture	c citrus f small fruits g grapes o pome fruits s stone fruits v vegetables		
R degraded natural vegetation intensively used, recovery likely to be protracted at best	R4 closed reedland Y4 closed sedgeland Y3 sedgeland Y2 open sedgeland	F forestry	e exotic softwood h hardwood x experimental		
		C conservation R recreation, tourism U urban fringe T urban M mineral extraction N no defined use	n nature w water		
C cultural – e.g. gardens, parks, orchards	F forest plantation W horticultural plantation P parkland P1 parkland over improved pastures or crops P(1) parkland over locally improved pastures O open parkland O1 open parkland over improved pasture or crops O(1) grass parkland over locally improved pastures I improved pastures or cropland				

Part Four

Quantitative Analysis of Data from Ecological Surveys

11 Data analysis and interpretation I: introduction and the Mann-Whitney U test

11.1 Introduction

Once the field data have been collected for a specific project, attention turns to the methods for the analysis of the data and the presentation of the results. Two very important points must be stressed immediately:

1 the data should have been collected with the techniques of analysis and presentation in mind. This emphasizes the importance of careful planning of the complete project. Data should not be gathered without prior thought about suitable methods for tabulation and graphical and statistical analyses;
2 in many cases the objective of a study is to test a hypothesis (Section 1.3) on some wider data set. Analysis of data collected for this purpose requires the use of inferential statistical methods and in this case the data must be collected to be representative of this more general context.

11.2 Inferential statistics

The statistical methods presented in the following chapters have been chosen for their simplicity and effectiveness in manipulating and displaying biological and environmental data. They have also been selected for their suitability for analysing the small quantities of field data usually generated by student projects. Many more sophisticated techniques exist. These are deliberately excluded either because of their complexity or because their use is only justified where large quantities of field data are involved.

Non-parametric statistical tests

All the methods of statistical analysis introduced in the next three chapters are non-parametric or 'distribution-free' in nature. Non-parametric methods are particularly suited to the types of data which are generated in student projects, since they make fewer assumptions about the data. The most important of these is that the data do not have to be normally distributed. The removal of the requirement for normality also means that non-parametric methods are well suited to small data sets.

Chapter 11 discusses the application of the Mann-Whitney U test to test for differences between two sets of observations – in this case, the floristic richness of hedgerows in Devon compared to those of Huntingdonshire. Chapter 12 discusses the use of the Chi-squared (χ^2) to investigate the relationship between a type of moth caterpillar and particular plants; and the grazing preference of sheep and cattle in upland pasture. Chapter 13 discusses the formulation of hypotheses, the use of scattergrams, rank correlation coefficients and regression by semi-averages to establish the nature of the statistical relationships between plant distributions and competitive and environmental controls. The use of a rank correlation matrix and correlation bondage diagram to study the relationship between topographic and pedological factors and vegetation down a soil catena is also discussed.

These chapters describe the use of inferential statistics. The purpose of inferential statistics is to test hypotheses. The nature of the hypothesis will vary from one problem to another, but the basic process of hypothesis generation is fairly standard and is explained below.

Setting up the null hypothesis

This is the first stage of any statistical analysis and states the hypothesis which is to be tested. This is the assumption which will be maintained unless the data provide significant evidence to discredit it. The null hypothesis is denoted symbolically as H_0.

As an example, we may wish to compare two samples of the numbers of plant species in hedgerows from different parts of the British Isles; the Counties of Devon and Huntingdonshire. The null hypothesis would be:

H_0: there is no difference in the number of plant species per unit length of hedgerow between the two samples of hedgerows from the two counties (the numbers of the plant species in the two sets of hedgerows are the same).

However, it is generally also necessary to state specifically the alternative hypothesis (H_1). In this case the alternative might be:

H_1: there is a difference in the number of plant species in the two samples of hedgerows per unit length of hedgerow in the two counties.

It is particularly important to specify the alternative hypothesis very precisely when departures from the null hypothesis of only one particular kind, are of interest.

Directional and non-directional testing

The alternative hypothesis H_1 may be directional or non-directional. A directional hypothesis (or one-sided hypothesis) is used when either only positive or negative differences are of interest in the study. For example, when an alternative biological or environmental hypothesis predicts that the mean of one sample would be greater (but not less) than another, then a directional alternative would be used. A non-directional (or two-sided) hypothesis would be used when both positive and negative differences are of equal importance in providing evidence with which to test the null hypothesis.

In the hedgerow example above, a preliminary null assumption H_0 is that both counties have the same mean number of species per unit length of hedgerow. An alternative 'environmental model' might not specify which county has the greater number of species per unit length of hedgerow and so the alternative hypothesis would be presented as two-sided i.e. non-directional.

Had the alternative environmental model been more specific (for example, that the conditions in the South-West encourages greater species diversity), then the null hypothesis might have been tested against the specific alternative: H_1 hedgerows in Devon contain more plant species per unit length of hedgerow than those in Huntingdonshire.

Once the null and particularly the alternative hypotheses have been stated, the next stage is to select the appropriate test statistic to be used in the study. This must be sensitive to departures from the null hypothesis towards the particular alternative tested for or against. The particular form of the test statistic and hence the test itself must vary from one situation to another. The final stage is to assess the statistical significance of the result of the test.

Significance levels and significance tables

For a given null hypothesis, the calculated value of the test statistic is compared with tables of critical values at specified significance levels. This comparison allows us to assess the degree of confidence with which we could reject the null hypothesis and is expressed as a probability or significance level. For example, if a calculated test statistic exceeds the critical value for a significance level of 0.05 then this means that values of the test statistic as large as, or larger than that calculated from the data, would occur by chance less than five times in one hundred if the null hypothesis were indeed true. In simpler words, if we were to reject the null hypothesis on the basis of this improbably large value of the test statistic, we would run a risk of less than 5 per cent of acting incorrectly. Rejecting a null hypothesis at a significance level of 0.01 means that we run a risk of less than 1 in 100 of erroneously rejecting a true null hypothesis, so we can say that the data provide evidence at the one per cent level of significance against the null hypothesis in favour of the alternative.

Levels of rejection

Normally a significance level is chosen prior to the test; this level is known as the level of rejection. The critical value for the test statistic is then found for that significance level from significance tables. If the value of the test statistic is greater than this critical level, then there is an even smaller probability than this specified probability level that the null hypothesis will be incorrectly rejected, though there are exceptions which we mention below. This means that the rejection of the null hypothesis can be justified in these probabilistic terms. In the hedgerow example, the null hypothesis – H_0 that there is no difference between the number of plant species per unit length of hedgerow in Devon and Huntingdonshire – is rejected (see below). The statistical analysis suggests the evidence of differences between the species richness of the hedgerows in the two counties is statistically significant; or (more loosely phrased) that there is a statistically significant difference between the two species diversities recorded.

However, if the calculated value of the test statistic is less than the critical value in the tables, then the probability that the null hypothesis will be incorrectly rejected is greater than the predetermined rejection level. In this case the rejection of H_0 is not warranted; the null hypothesis must be accepted. The statistical analysis has not demonstrated significantly indicative differences between the numbers of plant species per unit length of hedgerows in the two counties of the example.

The form of the significance test varies according to whether or not the alternative hypothesis was directional or non-directional. If this stated hypothesis was directional, then a one-tailed significance test is applied. This tests for significance in the stated direction of the difference, e.g. whether Devon has more plant species per unit length of hedgerow than Huntingdonshire. Where the hypothesis is non-directional, a two-tailed test is used. This tests for differences in either direction, e.g. either Devon hedgerows had more species than those of Huntingdonshire; or alternatively whether the hedgerows of Huntingdonshire had more species than those of Devon. Whether to use one- or two-tailed tests is determined by the practical situation involved, in this case the implications of the competing 'environmental' model. In the case of some commonly used statistical tests, it is more convenient to tabulate *lower* critical values. In these cases, the test results in rejection if the calculated value of the test statistic is *less than* the tabulated critical value. Care must be taken with such tests, and in particular, the Mann-Whitney U test. With this procedure the null hypothesis is rejected if the test statistic is less than the critical value. The null hypothesis is accepted if the test statistic is greater than the critical value. This is the reverse of the more usual principle.

11.3 Tabulation and checking of data

On returning from the field the plant, animal and/or environmental data which have been collected from notebooks or checksheets must be collated systematically. This might be on to a large master data sheet (even A1 size paper), or on to computer compatible data preparation forms. The particulars of the form and layout of this sheet will be determined primarily by the nature of the particular project and structure of the data; but delays and uncertainties at this stage encourage loss of data and in consequence loss of accuracy in the results.

11.4   The Mann-Whitney U test

In project work it is often necessary to test whether two samples of the same phenomenon are statistically different. Perhaps they may have been collected from different environments or under different sets of conditions. For example, the abundance of one species of plant may have been sampled on two different rock types, or the numbers of grazing animals may have been counted in samples from different vegetation types. Here the hypothesis that there is a statistically distinct difference between the abundance of the plant species in the two environments may be tested; or the aim could be to investigate whether there is a statistically significant difference in the numbers of grazing animals on the two different vegetation types. One test of particular use in this situation is the Mann-Whitney U test, which is used for independent samples. We illustrate this with a case study.

Case study: Variations in the plant species richness of hedgerows in Devon and Huntingdonshire, UK
Hedgerows are a very important wildlife reservoir within the agricultural landscapes of much of temperate Western Europe (Hooper 1970(a)). In many other parts of the world shelterbelts play the same role. They are important as habitats for both plants and animals, and in particular for bird and rodent species.

Habitat variations in hedgerows and shelterbelts
Figure 11.1 shows the diversity of habitats that occur across a mature hedgerow. Elton and Miller's description of habitat types (see Section 7.4) indicates there may be open ground habitats, either bare ground or close-cropped grassland in the fields on either side of the hedge. The area close to the hedge is usually composed of tall grass and herb vegetation of the field layer type which borders the scrub/thicket habitat of the hedgerow itself. The same pattern of variation is usually repeated on the other side, especially adjacent to roads or tracks. A ditch may sometimes be found which increases habitat diversity even further. This rapid variation in habitats across hedgerows is similar to that found at woodland margins. This led Elton and Miller to describe them as a special 'edge' type of wildlife habitat (Section 7.4).

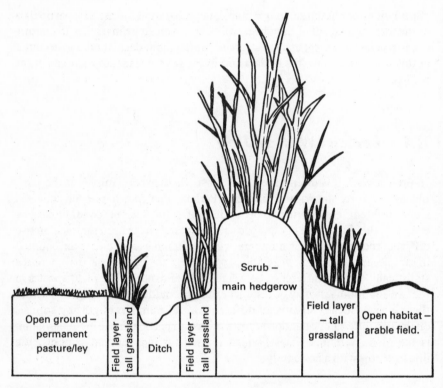

Figure 11.1   Variation of habitat types across a hedgerow or shelterbelt

Modern agriculture and hedgerows

Hedgerows are under considerable pressure from modern agricultural practices. The use of herbicides and pesticides has decreased their species diversity in many arable areas. Substantial lengths of hedgerow have been removed completely in order to provide larger fields which can be worked more economically by modern farm machinery. The effect of this has been to make the remaining hedgerows even more important as remnant habitats and to encourage study of the origins, structure and biological composition of hedgerows.

Many factors may influence the plant species diversity of a hedgerow. The age of the hedge has been shown to be important in determining shrub species diversity (Hooper 1970(b)); but local soil conditions, past and present hedgerow management practices and current agricultural activities are all important.

The present study

In the light of this information, a study was made of hedgerows in two markedly different agricultural environments in Britain. Samples of hedgerows were studied from Devon, where livestock farming is dominant and from

Huntingdonshire where agriculture is largely arable. Teather (1970) has shown that in Huntingdonshire some 90 per cent of the county's hedgerows were removed between 1946 and 1967. In contrast, Locke (1962) calculated that Devon had the greatest length of hedgerows of any county in Britain; no less than 86 per cent of the original field boundaries were still hedged.

The statistical test
One of the main aims of this study was to see whether the species diversity of higher plants was significantly different in the remaining hedgerows of Huntingdonshire compared with those in Devon.

It might be thought that the conditions in Devon, (well-established hedgerows, livestock farming with consequently less use of herbicides), would encourage species diversity to be greater per unit length of hedgerow than in the county of Huntingdonshire where arable farming is prevalent. It is therefore of interest to see whether the available data provide evidence in favour of this model against the null hypothesis that there is no appreciable difference between the sites. This environmental/biological framework determines the directional form of the alternative hypothesis in the statistical formulation.

The null hypothesis was constructed as:
H_0: the numbers of plant species in Devon hedges are not significantly different from those in Huntingdonshire.

The alternative directional hypothesis was:
H_1: plant species numbers for hedges in Devon are significantly higher than in Huntingdonshire.

The data to test this hypothesis were collected by sampling 32 m lengths of 15 different hedgerows in Huntingdonshire and 13 similar lengths in Devon. Within each length, the total numbers of species of higher plants were counted. The resulting data are presented in the first two columns of Table 11.1. The Mann-Whitney U-test was then used to test hypothesis (H_1). Note that the test can be applied to two samples with differing numbers of observations.

Calculation
To calculate the U-test statistic, the values for both sets of samples are ranked together on a continuous scale from lowest to highest. Where ties occur, the mean rank of all the scores involved in the tie is given to those observations (see Section 13.2). The rank values for each set of observations are then summed separately to give a value of:

$\sum r_1$ and $\sum r_2$

These values are then entered in the formulae shown under Table 11.1 for the calculation of U and U_1.

Table 11.1   *Analysis of numbers of higher plant species in hedgerows in Devon and Huntingdonshire using the Mann-Whitney U Test*

Number of higher plant species in a 32 m length of hedge		Rank n_1	Rank n_2
Devon (n_1)	*Huntingdon* (n_2)		
28	14	26	5
27	20	25	13.5
33	16	28	8.5
23	13	20	2.5
24	18	23	11
17	21	10	16
25	23	24	20
23	20	20	13.5
31	14	27	5
23	20	20	13.5
23	20	20	13.5
22	14	17	5
15	11	7	1
	16		8.5
	13		2.5
$n_1 = 13$	$n_2 = 15$	$\Sigma\, r_1 = 267$	$\Sigma\, r_2 = 139$

$$U = 13 \times 15 + \frac{13\,(13 + 1)}{2} - 267 = 19$$

$$U_1 = 13 \times 15 + \frac{15\,(15 + 1)}{2} - 139 = 176$$

$$U \text{ (lower value)} = 19$$

$$U = n_1 n_2 + \frac{n_1[n_1 + 1]}{2} - \Sigma\, r_1$$

$$U_1 = n_1 n_2 + \frac{n_2[n_2 + 1]}{2} - \Sigma\, r_2$$

where n_1 = the number of observations in the first sample
      n_2 = the number of observations in the second sample

The lower of these two values (U and U_1) is then taken to assess the significance of the difference between the two sets of samples. In this case, the direction of the test is known, because it has been hypothesized that the hedgerows in Devon have higher numbers of plant species than those in Huntingdonshire and so a one-tailed test is used.

In the example of Table 11.1, the value of U is 19 and U_1 is 176. The lower value is thus 19 and this value is then entered to the significance tables for Mann-Whitney U (Table 11.2) at an appropriate significance level for the given sample sizes. An important feature of this statistical test is that the greater the difference between the two sets of samples, the smaller will be the test statistic i.e. the lower value of U or U_1. Thus if the computed value is lower than the critical value in Table 11.2, the null hypothesis (H_0) is rejected for the given significance level. If the computed value is greater than the critical value, then the null hypothesis is accepted.

Using a significance level of 0.05 with sample sizes of $n_1 = 13$ and $n_2 = 15$, the critical value in the table for a one-tailed test is 61. Note that this is a one-tailed test, because the direction of the relationship is specified. The computed value of U is 19, which is very much less than the tabulated value. Thus the null hypothesis (H_0) is rejected, and the stated hypothesis (H_1), that there are significantly higher numbers of plant species in the hedgerows of Devon rather than Huntingdon, is accepted in preference.

One further word of caution is required on the Mann-Whitney U test. If the size of the samples exceeds 20, then Table 11.2 cannot be used. Instead either more extensive tables should be used or a reasonably accurate approximation which involves a different value called a z (zed) score is calculated; and a different set of significance tables is required (see Hammond and McCullagh 1978, p. 204).

Interpretation

Interpretation of this result is interesting and helps to emphasize the care required at this interpretative stage in data analysis. The proven difference in species richness could be due to a large number of factors (Pollard *et al.*, 1974). The Devon hedges may be growing on a different substrate, may be managed differently, have experienced successional processes to a greater or lesser degree depending upon their age and location, or have been planted using different species and techniques. The farming regimes are primarily arable in Huntingdonshire and livestock rearing in Devon. The diversity of surrounding habitats also varies between the two counties.

Consequently, although a difference has been shown to exist, the real reasons for the observed differences have not been discovered. This is an important point with which to conclude this section. The erecting of hypotheses, use of statistical analyses and the presentation of results is only a means to an end. Invariably once results have been examined and interpretations made, new ideas and hypotheses will emerge and further research suggested. Research and project work are often never ending processes of successive hypothesis generation and testing, which continuously throw up new ideas, awaiting the construction of further research projects.

Table 11.2 Critical values of U for the Mann-Whitney U test (Adapted from Siegel (1956); after Auble (1953).)

0.025 (one-tailed), 0.05 (teo-tailed)

n	1	2	3	4	5	6	7	8	9	10	11	12	13	14	15	16	17	18	19	20	n
2								0	0	0	0	1	1	1	1	1	2	2	2	2	2
3					0	1	1	2	2	3	3	4	4	5	5	6	6	7	7	8	3
4				0	1	2	3	4	4	5	6	7	8	9	10	11	11	12	13	13	4
5			0	1	2	3	5	6	7	8	9	11	12	13	14	15	17	18	19	20	5
6			1	2	3	5	6	8	10	11	13	14	16	17	19	21	22	24	25	27	6
7			1	3	5	6	8	10	12	14	16	18	20	22	24	26	28	30	32	34	7
8		0	2	4	6	8	10	13	15	17	19	22	24	26	29	31	34	36	38	41	8
9		0	2	4	7	10	12	15	17	20	23	26	28	31	34	37	39	42	45	48	9
10		0	3	5	8	11	14	17	20	23	26	29	33	36	39	42	45	48	52	55	10
11		0	3	6	9	13	16	19	23	26	30	33	37	40	44	47	51	55	58	62	11
12		1	4	7	11	14	18	22	26	29	33	37	41	45	49	53	57	61	65	69	12
13		1	4	8	12	16	20	24	28	33	37	41	45	50	54	59	63	67	72	76	13
14		1	5	9	13	17	22	26	31	36	40	45	50	55	59	64	69	74	78	83	14
15		1	5	10	14	19	24	29	34	39	44	49	54	59	64	70	75	80	85	90	15
16		1	6	11	15	21	26	31	37	42	47	53	59	64	70	75	81	86	92	98	16
17		2	6	11	17	22	28	34	39	45	51	57	63	69	75	81	87	93	99	105	17
18		2	7	12	18	24	30	36	42	48	55	61	67	74	80	86	93	99	106	112	18
19		2	7	13	19	25	32	38	45	52	58	65	72	78	85	92	99	106	113	119	19
20		2	8	13	20	27	34	41	48	55	62	69	76	83	90	98	105	112	119	127	20
n	1	2	3	4	5	6	7	8	9	10	11	12	13	14	15	16	17	18	19	20	n

0.05 (one-tailed), 0.10 (two-tailed)

n	1	2	3	4	5	6	7	8	9	10	11	12	13	14	15	16	17	18	19	20	n
1																			0	0	1
2					0	0	0	1	1	1	1	2	2	2	3	3	3	4	4	4	2
3			0	0	1	2	2	3	3	4	5	5	6	7	7	8	9	9	10	11	3
4			0	1	2	3	4	5	6	7	8	9	10	11	12	14	15	16	17	18	4
5		0	1	2	4	5	6	8	9	11	12	13	15	16	18	19	20	22	23	25	5
6		0	2	3	5	7	8	10	12	14	16	17	19	21	23	25	26	28	30	32	6
7		0	2	4	6	8	11	13	15	17	19	21	24	26	28	30	33	35	37	39	7
8		1	3	5	8	10	13	15	18	20	23	26	28	31	33	36	39	41	44	47	8
9		1	3	6	9	12	15	18	21	24	27	30	33	36	39	42	45	48	51	54	9
10		1	4	7	11	14	17	20	24	27	31	34	37	41	44	48	51	55	58	62	10
11		1	5	8	12	16	19	23	27	31	34	38	42	46	50	54	57	61	65	69	11
12		2	5	9	13	17	21	26	30	34	38	42	47	51	55	60	64	68	72	77	12
13		2	6	10	15	19	24	28	33	37	42	47	51	56	61	65	70	75	80	84	13
14		2	7	11	16	21	26	31	36	41	46	51	56	61	66	71	77	82	87	92	14
15		3	7	12	18	23	28	33	39	44	50	55	61	66	72	77	83	88	94	100	15
16		3	8	14	19	25	30	36	42	48	54	60	65	71	77	83	89	95	101	107	16
17		3	9	15	20	26	33	39	45	51	57	64	70	77	83	89	96	102	109	115	17
18		4	9	16	22	28	35	41	48	55	61	68	75	82	88	95	102	109	116	123	18
19		4	10	17	23	30	37	44	51	58	65	72	80	87	94	101	109	116	123	130	19
20		4	11	18	25	32	39	47	54	62	69	77	84	92	100	107	115	123	130	138	20
n	1	2	3	4	5	6	7	8	9	10	11	12	13	14	15	16	17	18	19	20	n

11.5   Projects

Classroom projects

General applications of Mann-Whitney U tests
There are many projects within this book where the use of the Mann-Whitney U test is appropriate at the analysis stage. Look through the projects at the end of other chapters and find examples. This will require the generation of hypotheses, statement of the null and alternative hypotheses and whether the test is to be directional or non-directional.

Field projects

Road verge studies
Road verges are man-modified habitats found in most parts of the world, and like hedgerows are often very interesting 'edge' habitats.

1   Using roads of a different type – dual carriageways, major class roads, minor roads and trackways, formulate hypotheses concerning species numbers bordering different road types.
2   Collect data relevant to your hypotheses – take great care sampling road margins. (Do not get run over!).
3   Use the Mann-Whitney U test on your data to test your hypotheses.
4   Present and discuss your results.
5   Identify environmental factors likely to cause variations in road verge conditions e.g. soil and rock type, verge management practices such as cutting, or the application of herbicides or winter salting.
6   Devise further projects centred on any of the hypotheses generated.

12 Data analysis and interpretation II: use of χ^2 to measure association and the χ^2 test

12.1 Animal distributions in relation to vegetation

Relationships between the distribution of animals and vegetation exist first because plants provide the food source for many animals and represent the base of the trophic pyramid (see Section 2.2); and second because plants and vegetation act as a habitat within which many organisms live, grow and reproduce. For these reasons, relationships may be found between particular animal and plant species or vegetation types and this forms the justification for habitat survey methods (see Section 7.10). This type of association can be investigated and analysed very simply using the chi-squared (χ^2) test. This test is often incorrectly termed the chi-square test.

Moth caterpillars and ragwort in a suburban garden

The relationship between the caterpillar stage of the cinnabar moth (*Tyria jacobaeae*) and ragwort (*Senecio jacobaea*) is well known (Dempster, 1971). The caterpillars of the cinnabar moth are easily identifiable by their striking black and yellow banding. This is a good example with which to demonstrate a technique for statistically assessing the association between an animal and a plant species. In Britain during May and early June, the caterpillars are frequently found grazing on ragwort plants. Often whole plants may be denuded of leaves over large areas. It is an interesting study to examine the extent to which the caterpillars are uniquely associated with the ragwort.

To discover this, and to test the hypothesis of positive association between the two species, the statistical chi-squared (χ^2) test on a 2 × 2 contingency table is used. A positive association results in caterpillars occurring with ragwort and absence of ragwort resulting in an absence of caterpillars. Ragwort tends to occur with the caterpillar. Similarly, if ragwort is absent, the caterpillar will be absent.

Field procedures

The sampling procedure involves laying out a large quadrat (10 m × 10 m) in an area where *Senecio jacobaea* is known to thrive in early summer. Many unattended back gardens or urban plots make excellent study sites in this respect. Within the 10 m × 10 m quadrat, individual plants of *Senecio* and other species may be examined, by taking random co-ordinates using random number tables (Table 5.4) to locate the sampling locations (see Section 5.5). The nearest individual plant to each random point is inspected, a note made of whether or not it is *Senecio* and whether or not the caterpillars of the cinnabar moth are found on the plant.

12.2 The chi-squared test (χ^2) of association

Once collected the data are partitioned into four categories as set out below:

a the number of individual ragwort plants with cinnabar moth caterpillars;
b the number of individual ragwort plants without cinnabar moth caterpillars;
c the number of individuals of other plant species with cinnabar moth caterpillars;
d the number of individuals of other plant species without cinnabar moth caterpillars.

These data are drawn up as a 2 × 2 contingency table as set out below, with a, b, c and d corresponding to the four categories above.

		Ragwort		
		Present	*Absent*	
Cinnabar moth caterpillars	*Present*	a	c	a + c
	Absent	b	d	b + d
		a + b	c + d	n

where n = total.

Note that the values for a, b, c, and d are known as frequencies and chi-squared may only be applied to such frequency data. They should not be applied to data which have been converted to percentages.

In this example taken from a suburban garden in Plymouth, Devon, 100 plants have been selected and examined using random number tables within the 10 m × 10 m quadrat. The actual data are inserted into the contingency table which now reads:

Ragwort

Cinnabar moth caterpillars		Present	Absent	
	Present	54	10	64
	Absent	13	23	36
		67	33	100

The hypothesis that no association exists between the ragwort and the cinnabar moth may be tested against the alternative that an association exists by applying the formula for chi-squared with Yates's correction for 2×2 contingency tables. This is:

$$\chi^2 = \frac{[|ad - bc| - 0.5n]^2 \times n}{[a + b] [c + d] [a + c] [b + d]}$$

Here $|ad - bc|$ means the modulus or the absolute or positive difference between ad and bc.

$$\chi^2 = \frac{[|(54 \times 23) - (13 \times 10)| - 50]^2 \times 100}{[64 \times 36 \times 67 \times 33]}$$

$$\chi^2 = \frac{1062^2 \times 100}{5094144}$$

$$\chi^2 = \frac{112784400}{5094144}$$

$$\chi^2 = 22.14$$

The computed value of chi-squared must be compared with critical values in tables to determine the significance of the pattern indicated. To determine the critical value of chi-squared, we need to know the appropriate degrees of freedom (df) and the level of rejection to be employed. Here the degrees of freedom are calculated according to the following formula:

df = [number of rows (r) in the contingency table minus 1] multiplied by [the number of columns (k) in the contingency table minus 1]
df = $[r - 1] \times [k - 1]$

In the Plymouth case:

df = $[2 - 1] \times [2 - 1]$
df = 1

If we select a level of rejection of 0.01 then the null hypothesis will only be incorrectly rejected 1 time in 100, hence we only run the risk of one chance in

Table 12.1 *The Chi-squared Test* (χ^2) (From Siegel (1956); after Fisher and Yates (1974).)

df	0.10	0.05	0.01	0.001
1	2.71	3.84	6.64	10.83
2	4.60	5.99	9.21	13.82
3	6.35	7.82	11.34	16.27
4	7.78	9.49	13.28	18.46
5	9.24	11.07	15.09	20.52
6	10.64	12.59	16.81	22.46
7	12.02	14.07	18.48	24.32
8	13.36	15.51	20.09	26.12
9	14.68	16.92	21.67	27.88
10	15.99	18.31	23.21	29.59
11	17.28	19.68	24.72	31.26
12	18.55	21.03	26.22	32.91
13	19.81	22.36	27.69	34.53
14	21.06	23.68	29.14	36.12
15	22.31	25.00	30.58	37.70
16	23.54	26.30	32.00	39.29
17	24.77	27.59	33.41	40.75
18	25.99	28.87	34.80	42.31
19	27.20	30.14	36.19	43.82
20	28.41	31.41	37.57	45.32
21	29.62	32.67	38.93	46.80
22	30.81	33.92	40.29	48.27
23	32.01	35.17	41.64	49.73
24	33.20	36.42	42.98	51.18
25	34.38	37.64	44.31	52.62
26	35.56	38.88	45.64	54.05
27	36.74	40.11	46.96	55.48
28	37.92	41.34	48.28	56.89
29	39.09	42.56	49.59	58.30
30	40.26	43.77	50.89	59.70

Notes: The critical values of chi-squared given above show the probability that the calculated value of χ^2 is the result of a chance distribution. The larger the value of χ^2, the smaller is the probability that H_0 is correct.

one hundred of falsely declaring there to be an association when in fact there is none.

Entering Table 12.1 for df = 1 and a level of rejection of 0.01, the critical value of chi-squared is 6.64. The calculated value for the Plymouth cinnabar moth data was 22.14. This value is greatly in excess of 6.64. Therefore the null hypothesis is rejected at the 0.01 level.

This is only part of the story however. What this analysis has actually shown is that the distribution of the cinnabar moths in relation to the ragwort plants is not independent. The strength of the relationship must now be determined and an assessment made on whether or not the relation is a positive one (that is, the moth caterpillars occur mainly on the ragwort plants) as predicted, or a negative one (the moth caterpillars occur mainly on other plants).

The chi contigency coefficient

Some authors advocate the use of a quantity termed the contingency coefficient to answer the first question above. However, it provides nothing extra to the chi-squared value from which it is derived. There is little general agreement on the formula for it; it is not widely used and not repeated here.

Are associations positive or negative?

In order to check whether an association is positive or negative, the expected frequency for the joint occurrences of the ragwort and the cinnabar moth must be calculated. The expected frequency of joint occurrences (a) can then be compared with the observed frequency to determine whether they are occurring more frequently (positive association) or less frequently (negative association) than expected. The expected value for this cell (a) is computed by multiplying the row total for that cell [a + c = 64] by the column total for that cell [a + b = 67], and dividing by the total frequency [100].

Hence the expected value for cell (a) is:

$$\frac{[64 \times 67]}{100} = 42.88$$

The observed value in this study [54] is therefore higher than the expected value [43]; there are more joint associations than we would expect. Consequently the association must be positive. Therefore cinnabar moths are shown to be preferentially associated with ragworts. If the ragwort is not present, then neither are the cinnabar moth caterpillars.

If the observed value for joint occurrences had been less than that for the expected frequency of joint occurrences, then the association would have been negative. This would have indicated that ragwort tended to grow without the cinnabar moths grazing upon them and the cinnabar moths would be found on other plant species, not the ragwort.

12.3 Caution

Care must be taken in the application of chi-squared to contingency tables. The expected frequencies for all four permutations of the contingency table can be worked out by multiplying the row total by the column total for a given permutation and dividing by the number of observations. It has been shown that if one of these expected values falls below 5, then the test in this form becomes less reliable, but more exact tests are available. It is worth checking for this effect.

The chi-squared test is only applicable to frequency data where the actual number of individuals are used and must not be employed where data are collected as, or transformed into, percentages. Inspection of the Plymouth study above will show that the data were not collected as percentages; the number of observations totalled 100, which is quite different.

Finally, the chi-squared test for independence in 2×2 contingency tables is also extremely valuable for examining the associations between two species of plant, or two species of animal. Some care is necessary in their application in this situation because the results may be greatly influenced by the size of quadrat or sampled area used to assess the vegetation cover or animal frequencies.

12.4 The use of chi-squared to investigate over-grazing of upland pastures

At a different scale, it is possible to study the relationships between large herbivores and specific pasture types. As an example, in 1969, a research project was undertaken on the distribution of grazing animals on the major vegetation types of Dartmoor, Devon. Serious concern had been expressed over the possibility of over-grazing of certain vegetation types, particularly the *Agrostis-Festuca* grasslands, many of which were partially invaded by bracken (*Pteridium aquilinum*).

Field methods

In the study by Sayer (1969), two approaches were used to count the numbers of sheep and cattle on the different vegetation types of Dartmoor. The first was to use colour aerial photography at a scale of 1:10,000. A series of photographic transects were repeated in March, June, August and November. These were studied, and the animal distributions observed were plotted onto 1:10,360 Ordnance Survey sheets. The distribution of the animals could then be related to major categories of vegetation derived from the map of Dartmoor vegetation later published by Ward, Jones and Manton (1972).

The second approach involved the ground survey of 750 acres of the southern part of the moor. Field observations of the animals, using binoculars, were plotted directly onto the 1:10,360 O S sheet. First, a belt transect 500 yards wide and 4 miles long was traversed on foot, eight times during the summer of 1969. (One complete transect took about 4 hours). Second, a smaller (74 acre) area was studied from a fixed observation point over a long period of time with the exact positions of the animals being recorded regularly.

Once the animal distributions had been plotted on the maps their positions could be correlated with the major vegetation types on the moor which had been mapped from the same aerial photography. In this way, each grazing animal could be related to one of six major vegetation categories, as shown in Table 12.2. This map also enabled the area of each vegetation category to be calculated.

Statistical analysis

From the data in Table 12.2 it is possible to test the hypothesis (H_0) that there is no difference in the number of grazing animals on the different vegetation types, against the alternative (H_1) that certain vegetation types are preferentially grazed. This is done by the use of the chi-squared statistic which is used to test an observed distribution, in this case, the numbers of grazing animals on specific vegetation types, against an expected random distribution. This random distribution would be represented by an equal distribution of the numbers of grazing animals across all vegetation types.

The observed number of grazing sheep in each vegetation type is shown in column 3 of Table 12.2. The total is 213. If the sheep were evenly distributed across all vegetation types, then the expected distribution of sheep could be calculated from the relative extents of the different vegetation types in the sample area (see column 1).

As an example, since 41 per cent of the study area was covered by upland bog, one would anticipate that 41 per cent of the total number of sheep would be on that vegetation type, if there were no grazing preferences. Since 213 sheep were observed in total, 87 would be expected to occur on upland bog (41 per cent of 213). The other expected values in column 4 are derived in exactly the same way.

The departure of the observed numbers of sheep from the expected values may be tested using the formula for the calculation of chi-squared:

$$\chi^2 = \sum_1^n \frac{[O - E]^2}{E}$$

where:
O = the observed frequency of sheep on a vegetation type;
E = the expected frequency of sheep on a vegetation type;
n = the number of vegetation types.

Using the data in columns 3 and 4 of Table 12.2, the chi-squared value for sheep grazing is calculated for each of the six vegetation types as shown at the top pf page 216.

Table 12.2 *Vegetation types, areas and numbers of grazing sheep and cattle in the Dartmoor Survey area* (Adapted from Sayer (1969).)

Vegetation Type	(1) Area (Acres)	(2) % area surveyed	(3) Observed number of sheep	(4) Expected number of sheep	(5) Observed number of cattle	(6) Expected number of cattle
1 Upland bog *Sphagnum + Nardus + Molinia*	337	41	15	87	18	17
2 Damp heather moor *Molinia + Erica + Nardus*	59	7	8	15	5	3
3 Short-cropped hill grassland	225	28	98	60	7	11
4 Bracken invaded grassland	82	10	63	21	0	4
5 Heathland with gorse	15	2	1	4	0	1
6 *Calluna/Molinia* moorland	100	12	28	26	11	5
Totals	818	100	213	213	41	41

Dominant species in each vegetation type

1 Upland bog: *Molinia caerulea, Erica tetralix, Eriophorum angustifolium, Sphagnum* spp.

2 Damp heather moor: *Erica tetralix, Molinia caerulea, Agrostis setacea (curtisii), Trichophorum cespitosum*

3 Short-cropped hill grassland: *Agrostis tenuis (capillaris), Festuca ovina, Galium saxatile, Luzula campestris, Sieglingia decumbens (Danthonia decumbens), Anthoxanthum odoratum*

4 Bracken invaded grassland: *Pteridium aquilinum, Agrostis tenuis (capillaris), Galium saxatile, Luzula campestris, Vaccinium myrtillus, Potentilla erecta*

5 Heathland with gorse: *Calluna vulgaris, Agrostis setacea (curtisii), Erica cinerea, Ulex gallii, Vaccinium myrtillus*

6 *Calluna/Molinia* moorland: *Calluna vulgaris, Molinia caerulea, Trichophorum cespitosum, Agrostis setacea (curtisii), Vaccinium myrtillus*

(Recent species name changes are given in brackets)

$$\chi^2 = \frac{[15-87]^2}{87} + \frac{[8-15]^2}{15} + \frac{[98-60]^2}{60} + \frac{[63-21]^2}{21} + \frac{[1-4]^2}{4} + \frac{[28-26]^2}{26}$$

$$= 59.59 + 3.27 + 24.07 + 84.00 + 2.25 + 0.15$$

$$= 173.33$$

In this case the degrees of freedom are determined as one less than the number of vegetation types:

$$n - 1 = 6 - 1$$
$$= 5$$

The study of the chi-squared tables (Table 12.1) shows that if the 0.01 significance level is selected, with 5 degrees of freedom, the chi-squared value has to exceed 15.09 to be significant. The calculated value of 173.33 is clearly well over this. Therefore the null hypothesis (H_0) that animals are distributed evenly over this part of Dartmoor is rejected. Examination of the observed values shows a marked concentration of sheep on the short-cropped grasslands which are dominated by *Agrostis tenuis* (*capillaris*) (Common bent) and *Festuca ovina* (Sheeps fescue). Further pressure on these pastures is indicated because it is these areas which have been extensively invaded by bracken, rendering them less available as pastures.

These results agree with those from a subjective scale of grazing preferences devised by Hunter (1958), for the different hill pasture types of Britain. This scale is presented for comparative purposes in Table 12.3. The evidence indicates that both sheep and cattle show marked preferences for the *Agrostis-Festuca* pasture, whereas areas of *Molinia* and *Nardus* are avoided. Sheep also prefer *Calluna* (heather) and dry *Agrostis-Festuca* areas. On the other hand, cattle tend to graze areas of *Agrostis-Festuca* with *Deschampsia* that have been flushed with nutrients by mineral-rich stream water or overland flow. Once again these data and conclusions are important both for the

Table 12.3 *Relative degree of preference of sheep and cattle for different types of vegetation (e.g. swards); 1 with Deschampsia caespitosa: 2 with D. flexuosa* (After Hunter (1958); reproduced with kind permission of the British Association for the Advancement of Science.)

	Agrostis/ Festuca	Molinia	Nardus	Calluna vulgaris	1 Flushed Agrostis/ Festuca	2 Dry Agrostis/ Festuca
Cattle	++++	++	+	+	+++	+
Sheep	++++	+	+	+++	+	+++

purposes of biological conservation and improving land use within, and the profitability of, upland farms.

12.5 Projects

Classroom project

The use of chi-squared has been demonstrated to show the uneven distribution of grazing sheep on major vegetation types on Dartmoor. In Table 12.2, the last column shows the number of cattle recorded on each of the same vegetation types.

1 Using the same principles as used in the sheep example, test the hypothesis that the distribution of cattle is not spread evenly across the vegetation types.
2 Comment on the grazing preferences of the cattle as opposed to the sheep.

Field projects

Relationships between plant/species
1 Within an area of semi-natural vegetation carry out a survey of plant species composition using quadrats and an appropriate sampling design.
2 Record species in at least 100 randomly located quadrats using presence/absence methods.
3 Draw up contingency tables and calculate chi-squared as a measure of association for pairs of plant species. Test for significance and determine if a significant association is positive or negative.
4 Look at the positive associations between species and attempt to recognize groupings of species which may represent plant communities or associations.

Plant species diversities
1 Design a project to determine whether plant species diversity per unit area (richness) is related to variation in rock type within your local area. This should involve the study of geological and soil maps (see Appendix), the use of quadrats to determine the numbers of species on a systematic basis and the use of χ^2 to see if a relationship exists between plant species numbers and different rock types.
2 What are the limitations of this type of study?

The same methods may be applied to plant species diversities under different management regimes e.g. different frequencies of burning, grazing, mowing.

13 Data analysis and interpretation III: correlation and regression using Spearman's rank correlation coefficient and semi-averages regression

13.1 Relationships between plant distributions and their environmental controls

Limiting factors

The discussion of limiting factors in Section 2.4 showed that the distribution of both plant and animal species and communities may be controlled by one or more environmental factors such as frequency of severe frost, available soil moisture, soil nutrient status, pollution, or by biotic pressures such as trampling or grazing. Although several factors may interact to control the distribution of a species, ultimately the nature of one factor, (the master limiting factor), will become critical and that species will no longer be able to survive.

Scattergrams

Data on the relationships between, for example, a plant species and selected environmental conditions may be collected by throwing a series of quadrats over the range of the species and recording species abundance and values of selected environmental variables as described in Section 5.5. The results of such surveys may first be presented as a scattergram, as illustrated in Figure 13.1.

Sagebrush in the Californian Desert
In the first example (Figure 13.2) the abundance of the Sagebrush (*Larrea divaricata*) is plotted against the mean rainfall figures for a number of sites in California (Woodell *et al.*, 1969). Density counts in 93 one metre square plots, were recorded and the results plotted against the mean precipitation in centimetres per year, for each plot. A clear linear trend of increased density of *Larrea divaricata* with increase in rainfall is demonstrated.

Inspection of the scattergram (Figure 13.2) indicates a single exception to this general relationship. The authors explained this exception as a conse-

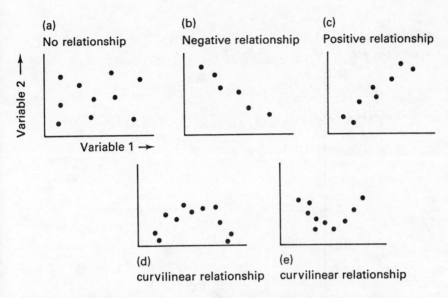

Figure 13.1 Scattergrams showing possible relationships between variables

quence of the particularly exposed conditions at this site. The relationship noted suggests a positive linear correlation between the two variables – sagebrush abundance and precipitation in the Californian desert.

Bracken fern on Yorkshire moorlands

A second example comes from the bracken fern (*Pteridium aquilinum*). The percentage cover of *Pteridium* was recorded in 28 one metre square quadrats on Millstone Grit in Yorkshire, England, together with estimates of soil moisture content. The latter were obtained by determining weight loss of soil samples after oven drying at 105°C for 24 hours. The values or scores for both variables – bracken cover and soil moisture content – were then plotted against each other. This produced the scattergram shown in Figure 13.3. In this case there is a clear trend of decreasing *Pteridium* cover with increasing soil moisture content. This indicates that a negative linear correlation may exist between the two variables.

Sometimes a scattergram will reveal no relationship between the two variables plotted (see Figure 13.1(a)). Other pairs of variables may show a strong negative (Figure 13.1(b)) or positive (Figure 13.1(c)) straight line (i.e. linear) relationship. Finally, some scattergrams will reveal non-linear (i.e. curved) relationships. It is very important to remember that introductory correlation and regression methods may only be used to define or clarify linear relationships as shown in Figures 13.1(b) and (c). Other statistical methods for dealing with curvilinear relationships are beyond the scope of this book.

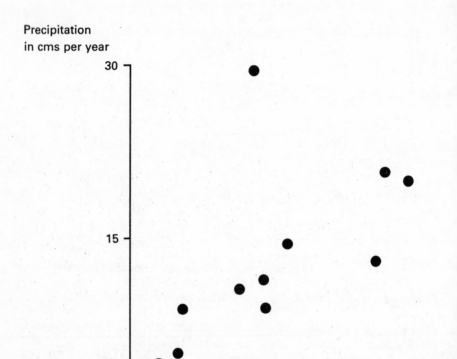

Figure 13.2 The density of *Larrea divaricata* stands in relation to precipitation at a number of sites (Redrawn from Woodell *et al.* (1969), with kind permission of the Journal of Ecology and Blackwell Scientific Publications Ltd.)

13.2 Spearman's rank correlation coefficient

The degree of correlation between *Pteridium* cover and soil moisture content may be determined by calculating a Spearman's rank correlation coefficient. First the hypotheses must be formulated.

Visual inspection of the scattergram suggests this relationship exists; but is it statistically significant? As a null hypothesis we take H_0: there is no correlation between *Pteridium* cover and soil moisture content.

If we are investigating an alternative biological model which predicts that there would be a negative correlation, then we test against the one-sided alternative H_1 of there being a negative correlation between *Pteridium* cover and soil moisture (a directional hypothesis using a one-tailed test).

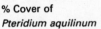

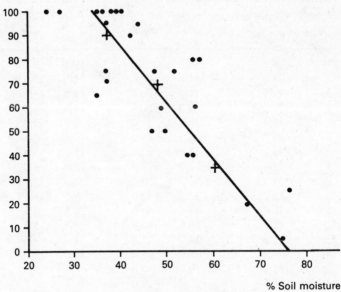

% Soil moisture

Figure 13.3   Semi-averages regression line fitted to a scattergram of percentage cover of bracken (*Pteridium aquilinum*) against percentage soil moisture content

Procedure

The method is set out in Table 13.1.

1   The two variables are ranked separately and consistently (i.e. always from high to low; or alternatively always from low to high values) and each observation is represented by a ranked value as shown in Table 13.1.

2   The arithmetic difference between each pair of ranked values is calculated and squared (see Table 13.1).

3   The sum of the squared differences is then computed (d^2). Where two observations have the same value, they are said to be 'tied'. In this case, the scores that the values would have taken are averaged, and the average is given to all those observations with the same value. As an example, in the soil moisture data there are two scores of 36.7 (see Table 13.1). These would be ranked as 6th and 7th. Thus those rank values are summed [6 + 7 = 13], and then averaged [13/2 = 6.5]. This average value of 6.5 is the final value for both original scores.

With the *Pteridium* data there are three quadrats with 75 per cent cover values. These would be ranked 13th, 14th, and 15th. When these three ranks are summed [13 + 14 + 15 = 42], and then averaged [42/3 = 14], the average rank of 14th is the final rank for all three scores of 75 per cent cover.

Table 13.1 *Calculation of the Spearman's rank correlation coefficient for the soil moisture content and Pteridium % cover data*

Quadrat	% soil moisture	Rank	% Pteridium cover	Rank	Difference in rank (d)	Difference^2
1	36.7	6.5	95.0	19.5	13.0	169.00
2	23.9	1.0	100.0	24.5	23.5	552.25
3	39.9	11.0	100.0	24.5	13.5	182.25
4	37.5	9.0	100.0	24.5	15.5	240.25
5	67.4	25.0	20.0	3.0	22.0	484.00
6	36.7	6.5	75.0	14.0	7.5	56.25
7	43.7	13.0	95.0	19.5	6.5	42.25
8	37.1	8.0	70.0	12.0	4.0	16.00
9	25.8	2.0	100.0	24.5	22.5	506.25
10	55.2	20.0	40.0	6.5	13.5	182.25
11	55.6	21.0	80.0	16.5	4.5	20.25
12	48.7	16.0	60.0	10.5	5.5	30.25
13	51.6	18.0	75.0	14.0	4.0	16.00
14	47.2	15.0	75.0	14.0	1.0	1.00
15	54.8	19.0	40.0	6.5	12.5	156.25
16	56.6	23.0	80.0	16.5	6.5	42.25
17	65.3	24.0	35.0	5.0	19.0	361.00
18	69.5	26.0	1.0	1.0	25.0	625.00
19	74.6	27.0	5.0	2.0	25.0	625.00
20	76.3	28.0	25.0	4.0	24.0	576.00
21	55.9	22.0	60.0	10.5	11.5	132.25
22	35.0	3.0	100.0	24.5	21.5	462.25
23	38.2	10.0	100.0	24.5	14.5	210.25
24	46.7	14.0	50.0	8.5	5.5	30.25
25	49.3	17.0	50.0	8.5	8.5	72.25
26	35.2	4.0	100.0	24.5	20.5	420.25
27	35.5	5.0	100.0	24.5	19.5	380.25
28	42.1	12.0	90.0	18.0	6.0	36.00

$$\Sigma\ d^2\ =\ 6627.5$$

4   The differences are then entered into the following formula:

$$r_s = 1 - \frac{6\sum d^2}{n^3 - n}$$

where:

r_s = Spearman's rank correlation coefficient;
$\sum d^2$ = sum of the squared differences (see Table 13.1);
n = number of pairs of observations.

Thus:

$$r_s = 1 - \frac{6 \times 6627.5}{28^3 - 28}$$

$$r_s = 1 - \frac{39765}{21924}$$

$$r_s = 1 - 1.8137$$

$$= -0.81.$$

The value of r_s obtained has a negative value indicating a negative correlation. If a positive value had been obtained for r_s then a positive correlation would have been implied. The value of -0.81 suggests a strong negative correlation. The maximum values obtainable for r_s are -1 and $+1$; indicating perfect negative and positive correlations respectively. An r_s value of 0 indicates no correlation.

13.3 A statistical significance test using the Student's t distribution

This value of r_s must now be tested for its statistical significance, because although we have evidence of a negative correlation in this case, there is no idea as yet, whether the result might have been obtained by chance, or whether it represents a genuine relationship between the two variables; *Pteridium* cover and soil moisture. This point can be clarified by establishing another null hypothesis and testing it with another well known statistical method – the Student's t test.

The Student's t test

The null hypothesis (H_0) is that no correlation exists between the two variables. The Student's t statistic for significance testing is calculated using the following formula:

$$t = r_s \sqrt{\frac{n - 2}{1 - r_s^2}}$$

For the present example, this gives:

$$t = -0.81 \sqrt{\frac{28 - 2}{1 - 0.81^2}}$$

$$= -7.04$$

To assess the significance of the calculated value of t, it is compared with critical values of Student's t (given in Table 13.2), at a chosen level for what are termed

Table 13.2　*Student's t distribution: Table of values of t corresponding to specified one and two-tailed probabilities and degrees of freedom* (From Alder and Roessler (1964); after Fisher and Yates (1974).)

Degrees of freedom	One-tailed p = 0.05 p¹ = 95%	Two-tailed p = 0.05 p¹ = 95%	One-tailed p = 0.01 p¹ = 99%	Two-tailed p = 0.01 p¹ = 99%
1	6.31	12.71	31.82	63.66
2	2.92	4.30	6.97	9.93
3	2.35	3.18	4.54	5.84
4	2.13	2.78	3.75	4.60
5	2.02	2.57	3.37	4.03
6	1.94	2.45	3.14	3.71
7	1.90	2.37	3.00	3.50
8	1.86	2.31	2.90	3.36
9	1.83	2.26	2.82	3.25
10	1.81	2.23	2.76	3.17
11	1.80	2.20	2.72	3.11
12	1.78	2.18	2.68	3.06
13	1.77	2.16	2.65	3.01
14	1.76	2.15	2.62	2.98
15	1.75	2.13	2.60	2.95
16	1.75	2.12	2.58	2.92
17	1.74	2.11	2.57	2.90
18	1.73	2.10	2.55	2.88
19	1.73	2.09	2.54	2.86
20	1.73	2.09	2.53	2.85
21	1.72	2.08	2.52	2.83
22	1.72	2.07	2.51	2.82
23	1.71	2.07	2.50	2.81
24	1.71	2.06	2.49	2.80
25	1.71	2.06	2.49	2.79
26	1.71	2.06	2.48	2.78
27	1.70	2.05	2.47	2.77
28	1.70	2.05	2.47	2.76
29	1.70	2.05	2.46	2.76
30	1.70	2.04	2.46	2.75
40	1.68	2.00	2.42	2.70
60	1.67	2.00	2.39	2.66

Notes: p is the probability of a value being more extreme than t. p¹ is the probability (expressed as a percentage) of a value being less than t.

the appropriate number of degrees of freedom. In the case of correlation coefficients, the degrees of freedom are equal to [n − 2]; in this example this is 28 − 2, which equals 26. A significance level of either 0.05 or 0.01 may be

Table 13.3 *Critical values of Spearman's rank correlation coefficient* (r_s) (After Siegel (1956) and Olds (1938 and 1934).)

	One-tailed test	
n	*0.05*	*0.01*
4	1.00	
5	0.900	1.00
6	0.829	0.943
7	0.714	0.893
8	0.643	0.833
9	0.600	0.783
10	0.564	0.746
12	0.506	0.712
14	0.456	0.645
16	0.425	0.601
18	0.399	0.564
20	0.377	0.534
22	0.359	0.508
24	0.343	0.485
26	0.329	0.465
28	0.317	0.448
30	0.306	0.432

selected. In this case, a more rigorous approach will be adopted and the 0.01 significance level will be taken. Thus the result will only be accepted as significant if it could only occur by chance in less than one case in one hundred.

A one-tailed test is applied here because it was hypothesized that the direction of the relationship is negative, that is, as soil moisture content decreases, so *Pteridium* cover increases.

For a one-tailed test at a significance level of 0.01 and with 26 degrees of freedom, the critical value of Student's t obtained from the table (Table 13.3) is 2.48. The calculated value of t in the example was 7.04; clearly well in excess of 2.48. Consequently, following the argument in Chapter 11 on significance tests, the null hypothesis that no relationship exists between the two variables of *Pteridium* cover and soil moisture is rejected. The result is impressive in that the choice of significance level indicates we might only expect the result to occur as a result of chance less than one time in a hundred. The hypothesis (H_1) of a negative correlation between percentage soil moisture and percentage cover of *Pteridium* is therefore accepted.

Unwarranted conclusions?

The use of rank correlation in this example suggests that the amount of bracken cover in the study area is a function of the soil moisture content. From this it might be inferred that *Pteridium* is sensitive to soil moisture conditions; or that it does not like wet conditions.

From this reasoning, further predictions could be made. For example, the installation of land drains on badly drained hillslopes adjacent to land where bracken occurs might promote the spread of the bracken, leading to an unexpected loss of the good pasture, beneath this rather unpalatable fern, thus defeating the objective of field drain installation. However, considerable caution is needed in making such predictions. The actual explanation of why the soil moisture/bracken cover relationship was found may be more complicated than first assumed. Simply because a pattern showing the response of bracken to soil moisture has been observed, it does not mean that soil moisture content is the only, or even the most important, variable controlling bracken distribution. For example, it might be argued (possibly somewhat deviously) that the *Pteridium* is responsible for the lower soil moisture content, rather than the other way around: the large fronds of bracken transpiring rapidly and therefore effectively draining the soil more rapidly. The role of grazing animals must be considered. Bracken is often prevented from growing and hence suppressed by the trampling effects of the large hooves of cattle and horses early in the growing season. Economic pressures encouraged many hill farmers to switch to sheep from cattle in the 1970s. Sheep tend to avoid the wetter ground, if possible. They do not graze the bracken, neither do their small feet have any significant trampling impact.

Consequently, care must always be taken when assuming a causal relationship exists merely because our quantitative studies have yielded a statistically significant correlation between two variables.

Correction factors for studies with many tied values

A further point concerning Spearman's rank correlation coefficient relates to the presence of tied values. If large numbers of tied scores occur in a calculation, the correlation coefficient will tend to be overestimated. A correction factor can be included, but the effect of the ties is usually too small to be important. In the *Pteridium*-soil moisture example there are a number of tied ranks, particularly in the *Pteridium* data. However, the correlation is very highly significant and the correction factor makes little difference to the coefficient value. Nevertheless, if a student finds very large numbers of ties and the results are only just significant, it is worthwhile incorporating the correction factor which is described in Siegel (1956, p. 206).

Small samples

Where there are less than 10 pairs of observations in a calculation, then the Student's t distribution cannot be used to test significance. Instead, the values in Table 13.3 must be used. In this table an appropriate significance level (for example 0.05 or 0.01) is chosen and the calculated value of r_s is examined for a given value of n (the number of pairs of observations). If the calculated r_s value exceeds the critical value for a given size of n at the selected significance level, then the result is significant. The null hypothesis (H_0) would be rejected. The stated alternative hypothesis (H_1), that there is a relationship, would be accepted.

The rank correlation coefficient has been used in this example to demonstrate a relationship between the abundance of a particular species and an environmental factor. The method is applicable in a very wide variety of other situations. For example, the relationship between two environmental factors, or between a different environmental control and another plant or animal species, or the relationship between two species, may be studied in the same manner as described above.

13.4   Regression by semi-averages

The data on soil moisture content and *Pteridium aquilinum* cover shown in Figure 13.3 display a clear statistical linear relationship which has been further demonstrated by the highly significant rank correlation coefficient between them. Although the scattergram has been plotted, it is not possible to see precisely the nature of the linear trend, nor can an attempt be made to precisely predict variation in *Pteridium* cover from the soil moisture data. In order to do this, it is necessary to find the line of 'best fit' through the scatter of points. This may be achieved through the process known as regression. It could be attempted by eye. For biological management purposes, a more reliable, quantitative approach is required.

One of the simplest methods is to calculate a regression line by the method of semi-averages. This greatly reduces the element of guesswork in fitting a regression line, although it does not eliminate the subjective element completely.

Dependent and independent variables

When examining two variables, the aim is often to seek to define a *causal relationship* between them. Neither correlation nor regression tells us this; they only show us if a significant association exists between two sets of numbers or observations. Instead, common sense must be used in interpretation. In this instance, it is probable that the percentage of soil moisture may determine the cover and vigour of *Pteridium* and not vice versa. Where it is reasonably clear which variable is causing change in the other, then the terms dependent and independent variables may be assigned to them. Here *Pteridium* is believed to be directly or indirectly dependent on soil moisture and is therefore considered to be the dependent variable. Soil moisture is consequently the independent variable. In scattergrams, the independent variable (x) is always plotted as the abscissa (horizontally). The dependent variable (y) is plotted as the ordinate (vertically).

Calculation of semi-averages

The significance and construction of a semi-averages regression line is most easily understood by inspecting Figure 13.3, which shows such a regression line plotted for the soil moisture – *Pteridium* data from Yorkshire.

Procedure

The following procedure is used to calculate the position of the semi-averages regression line.

1 Calculate the overall mean (averages) of both variables.
Overall mean of x (independent variable) (% soil moisture) = 47.9
Overall mean of y (dependent variable) (% cover *Pteridium*) = 68.6

2 Calculate the first semi-average.
x co-ordinate = mean of all values below the overall mean of x
15 values are below the mean of x; their average = 37.4.
y co-ordinate = mean of all values above the overall mean of y
17 values are above the mean of y; their average = 90.3.

3 Calculate the second semi-average.
x co-ordinate = the mean of all values above the overall mean of x
13 values are above the mean of x; their average = 60.1
y co-ordinate = mean of all values below the overall mean of y
11 values are below the mean of y; their average = 35.1.

These three sets of co-ordinates (x:y – 47.9:68.6; 37.4:90.3; and 60.1:35.1) for the overall mean, and the two for the upper and lower semi-averages are then plotted on the scattergram, together with the original data points (Figure 13.3).

These three points should very nearly form a straight line. The 'best-fit' semi-averages regression line can then be plotted with a ruler, by eye, with a high degree of accuracy, although this line need not actually join the points.

The method has the advantage of being fast and of not requiring a large amount of computation. However, it must be understood that more sophisticated and complex methods for regression are available, but are beyond the scope of this chapter.

If curvilinear relationships are found when plotting scattergrams (as illustrated in Figure 13.1(d) and (e)), then no simple techniques for fitting 'best-fit' regression lines are available. The simple answer is to draw in a line by eye. When doing this try to ensure that the points occur as frequently above the graph line as below it, for all sections of the line.

13.5 The use of a rank correlation matrix and correlation bondage diagram

Where data have been collected on several properties of vegetation and also a range of environmental factors, to produce 10 or 12 variables, rank correlation offers a useful means of analysing the data. The rank correlation coefficients calculated may be used to construct a rank correlation matrix, from which a correlation bondage diagram may be constructed. This latter diagram may indicate important ecological relationships underlying the patterns observed in the field. A single individual faced with the number of calculations necessary to deal with this data set would be justified in resorting to a computer to ensure accuracy as well as save time. In the following example, the field data were collected by a group of 25 students. The necessary calculations were produced by groups of two students with a pocket calculator, each determining the rank correlation coefficients for just two or three pairs of variables. The correlation matrix was constructed on a blackboard.

Case study: The inter-relationships between tree density, tree species diversity, and ground flora diversity down a soil catena
Lady Spring Wood occurs on the steep, west-facing slope of the River Sheaf in Sheffield, West Yorkshire. It is a well developed deciduous woodland, dominated by *Quercus* spp., *Acer pseudoplatanus*, *Ulmus* spp., and *Betula* spp., (oaks, sycamore, elm and birch). The underlying soils are developed on solifluccted shales and sandstones. Solifluction is a general term for the processes of slippage and flow downhill of soils under the influence of repeated freeze-and-thaw in present, and in the Sheffield case, past sub-arctic climates. The soils exhibit marked catenary relationships described below and illustrated in Figure 13.4. At the top of the slope are deep brown podsolic soils, which become shallower on the steeper middle parts of the hillslope. At the base of the slope, very waterlogged conditions occur, resulting in gleyed brown earths. The steepest slopes are thought to give rise to the greatest downslope movement of material and soil by soil erosion. They were also suspected to give rise to the greatest spacing between trees as a consequence of this, and consequently might have received relatively larger levels of light penetration through the tree leaf canopy in summer months when the canopy was fully open. It was further believed that these higher light levels would give rise to a distinctive and more diverse herb and ground flora on these steeper middle slopes.

In order to test this inter-related group of hypotheses derived from casual field observations, the properties of vegetation and soils were sampled along three transects down the hillslope, giving a total of 16 quadrats.

Within each quadrat the following variables were recorded on checksheets.

1 The percentage cover of all the higher plant species;
2 The girth of the five nearest trees;

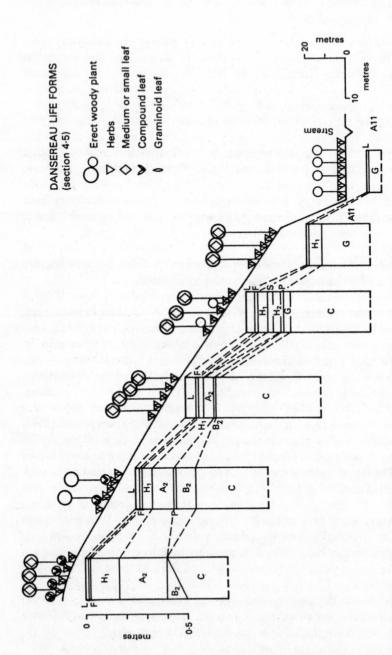

Figure 13.4   Downslope changes in soil and vegetational characteristics at Lady Spring Woods, Sheffield, England Key to soil horizons: **L** undecomposed litter; **H** humus layer; **H₂** second humus layer above sand lens; **A₂** eluviated horizon; **B₂** illuviated horizon; **C** little altered parent material (Carboniferous sandstones and relict periglacial drift); **G** mottled gley horizon; **S** sand lens; **All** alluvium.

3 The density of the trees, measured by taking the distance from the centre of the quadrat to the nearest five trees;
4 The number of tree species in the nearest 10 trees;
5 The number of ground flora species, as a measure of species richness;
6 The depth of the soil, defined as the surface of the C horizon;
7 The depth of the humus horizon;
8 The pH of the humus horizon;
9 The distance of the quadrat downslope from a fixed reference point;
10 The slope angle in the centre of each quadrat.

The soil variables and features were measured as described in Briggs (1977). The soil profile variations down one transect were presented as a soil and vegetation catena diagram which is shown as Figure 13.4. Changes in soil depth and horizonation are clearly revealed, and the use of simplified Dansereau symbols (see Section 4.5) provides a useful vegetation profile diagram.

Two species – the grass *Deschampsia cespitosa* (Tufted hair grass) and *Pteridium aquilinum* (Bracken) were separated from the percentage cover values and used as two separate variables. This was because they had the highest average cover values and hence were completely dominant in the ground flora. Rank correlation coefficients were calculated for all pairs of the variables described above. The pattern of significant correlations between these variables was then used to test the hypotheses concerning position on the slope, tree density, light intensity and ground flora characteristics.

The correlation matrix
Table 13.4 shows the rank correlation matrix for the 11 variables investigated; because the statistical analysis was worked by hand, not every permutation of variables was calculated. However, the most important correlations were tested for their statistical significance, and those which were found to be significant at the 0.01 level have been enclosed in a black box on the matrix.

The correlation bondage diagram
The correlation bondage diagram was constructed from this matrix as illustrated in Figure 13.5. The name of each variable has been placed inside a box on a separate diagram and connecting lines drawn to other boxes where the two variables had significant intercorrelations. Positive correlations are indicated by solid lines, negative correlations are shown by dashed lines.

Although not all possible combinations of correlations between the 11 variables were calculated, the interpretation of the results shows the hypothesized explanatory model above to be in need of some revision.

Interpretation – slope angle, soil, plant diversity and tree spacing
Slope angle is the most highly correlated variable and is negatively correlated at

Table 13.4   *A half-matrix showing correlations between variables measured down a soil catena at Lady Spring Woods, Sheffield, England*

	Pteridium cover	Deschampsia cover	Tree girth	Tree density	Tree diversity	Diversity ground flora	Soil depth	Humus depth	Humus pH	Distance downslope	Slope angle
Slope angle	+0.65	−0.85	−0.15	−0.13	−0.68	−0.78	−0.68	−0.88	−0.98	XX	XX
Distance downslope	−0.75	−0.48	−0.50	+0.59	+0.30	0.00	+0.58	−0.30	−0.10	XX	
Humus pH	−0.65	−0.88	−0.10	+0.60	+0.80	+0.68	XX	XX	XX		
Humus depth	+0.15	−0.88	XX	XX	XX	+0.58	XX	XX			
Soil depth	−0.15	−0.54	XX	XX	XX	+0.88	XX				
Diversity of ground flora	−0.65	−0.80	−0.13	−0.23	+0.95	XX					
Tree diversity	+0.20	−0.88	XX	XX	XX						
Tree density	−0.80	+0.13	XX	XX							
Tree girth	−0.70	+0.38	XX								
Deschampia cover	XX	XX									
Pteridium cover	XX										

Key

XX   rank correlation coefficient not calculated.

+ 0.6   rank correlation coefficient significant at 0.01 level ($r_s > 0.601$; one tailed test; n = 16)

the 0.01 significance level with humus pH and soil depth. This indicates that as the slope increases, so both humus depth and pH decreases. Soil erosion may be the cause, steeper slopes generating more rapid run-off and slope failure, hence removing soil, and preferentially exposing the rather acid bedrocks in the area.

Significant negative correlations also exist between slope angle, the number of tree and ground flora species, and the cover of the grass *Deschampsia cespitosa*. Thus it is gentler, not steeper slopes that are associated with greater species

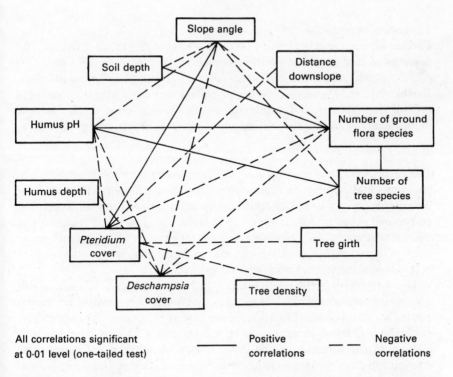

All correlations significant
at 0·01 level (one-tailed test) —— Positive correlations — — Negative correlations

Figure 13.5 A correlation bondage diagram for the vegetation and soil variables at Lady Spring Woods

richness and *Deschampsia*. Steeper slopes are also seen to be positively associated with *Pteridium*; the better drainage factor discussed previously may be at work.

It is interesting that tree density is not significantly correlated with slope angle, nor is tree density correlated with the number of ground flora species. Thus the original hypothesis of a direct relationship between tree diversity and ground flora is not acceptable in this instance. Instead, the diversity of the ground flora is positively correlated with the pH of the humus, and negatively with slope angle. The plant species richness is related to soil factors.

These results together with the correlations of other soil variables with slope angle suggest:

1 that the impoverished nature of the shallow mid-slope soils is affecting ground flora diversity and that the highest diversities are found on less steep slopes where the leaching of soil nutrients is least;
2 and/or that the ground flora is affecting the type and pH of the soil humus.

These analyses indicate that soil-plant relationships appear to be more important than light in determining species diversity in this instance.

The vegetation variables

Further interesting relationships occur between the vegetation variables. The diversity of tree species is positively related to ground flora diversity. This could suggest greater tree diversity encourages diversity in the ground flora. *Deschampsia* and *Pteridium* cover are negatively correlated with the number of ground flora species, indicating that where present, these species achieve dominance over and suppress other potential competitors in the ground flora.

Discussion

The establishment of networks of intercorrelations between vegetation and environmental variables in this manner is a very useful method for studying the complexities of soil-plant relationships and seeking to explain vegetation and animal distributions. However, this study also shows how the testing of what may seem to be a relatively straight-forward set of hypotheses on an unexceptional area of woodland, may lead to the generation of a totally new set of ideas requiring further work to develop and test.

The complexity of the relationships which have emerged even within this quite small data set clearly demonstrates that much more detailed research is needed on this woodland to evaluate our new ideas. One such study project might be a detailed investigation of soil chemistry and nutrient levels in relation to the diversity of the ground flora. Another is the competitive relationship that appears to exist between *Pteridium* and *Deschampsia*, and between these two and the other ground flora species. When properly verified, these relationships might be used to attempt to improve the diversity of the ground flora in the woodland.

13.6 Projects

Classroom project

Study of soil-slope relationships using rank correlation coefficients, scattergrams, and semi-averages regression lines.

Procedure

1   Examine the environmental data in Table 13.5.

2   Formulate hypotheses about the possible inter-relationships between: (a) soil/peat depth; (b) slope angle; (c) percentage soil moisture; (d) pH.

3   Draw scattergrams for those inter-relationships which you believe would be worth studying.

4   Using Spearman's rank correlation coefficient, test your hypotheses by calculating the coefficient between those variables for which you have drawn scattergrams.

5   For those relationships which have the highest significant correlations, calculate and plot the semi-averages regression line on the scattergram.

6 Discuss your results and formulate any new hypotheses which would be worth testing.

Field projects

Vegetation-soil-slope relationships
Using the soil–slope relationship study as a model, devise and carry out a project to examine soil/vegetation/slope relationships on a hillslope covered by natural or semi-natural vegetation in your local area.

Species-area relationships
Study the project on species-area relationships involving the use of Spearman's

Table 13.5 *Environmental data collected from 25 quadrats up a hillslope on Gutter Tor, Dartmoor, South-West Endland*

Quadrat number	Number of species	Soil peat depth (cms)	Slope angle (degrees)	% soil moisture	pH
1	6	54	3	95	4.8
2	5	42	1	110	4.6
3	5	12	20	34	3.8
4	7	15	6	75	3.9
5	5	16	18	55	4.0
6	5	11	10	52	3.7
7	5	9	9	51	3.9
8	5	15	20	31	3.7
9	8	40	4	95	4.1
10	8	38	3	76	4.2
11	7	31	4	67	3.9
12	8	25	4	105	4.2
13	7	17	15	35	4.2
14	6	10	12	43	4.5
15	4	14	25	15	4.3
16	6	15	9	50	3.7
17	5	23	10	52	4.2
18	8	18	15	36	4.0
19	6	20	17	42	4.2
20	6	10	5	43	3.9
21	4	8	7	50	3.6
22	6	16	2	72	4.0
23	3	27	12	33	4.1
24	6	13	13	21	4.2
25	8	55	1	112	5.2

rank correlation coefficient and semi-averages regression described in Section 13.3 and 13.4. Using this project as a model, design and carry out a similar one to study the relationships between plant species numbers and area in:

1 abandoned and derelict land, and
2 parks of different sizes and types, within your local area.

Part Five

Pollution, Conservation and Environmental Management

14 Pollution and environmental monitoring

14.1 The pollution problem

Environmental pollution is now receiving a great deal of attention. It is of increasing concern not just to biologists and ecologists, but also to many other professions including economists, planners, geographers and politicians. Pollution is a consequence of attitudes of mind, overpopulation and technological developments. Pollution is not a new problem, neither is it one which can be ignored. The nature and effects of pollution are variable, often being insidious rather than dramatic. Frequently people adopt, or are encouraged to adopt, a complacent or 'optimistic' attitude to the problem. It is argued that pollution is a necessary and acceptable consequence of an important industrial or agricultural process, or that it is essential to the maintenance of employment. A common, but debatable assumption, is that in 'good time' the appropriate techniques will be developed to reduce or eliminate the problem.

Many questions arise. What are the long term effects of pollution? Are projected trends of pollution concentrations reliable? What is the biological significance of the synergistic behaviour of pollutants. What will happen when high quality, low sulphur, coal and oil supplies become exhausted and the industrialized countries start to power their industries with the poorer, sulphur-rich fuels? What will happen when more countries in the developing world industrialize? How safe and secure are industrial processes which involve the use of noxious or toxic substances? How representative of field conditions are the laboratory experiments from which much of our data on the biological significance of pollutants are derived? The solutions to pollution problems are many and varied and lie in the various fields of education, improved production processes, improved pollution control technology, the production of pesticides and herbicides more akin to substances already in the environment, major changes in attitudes, political activity, legislation and litigation.

This chapter is concerned with demonstrating simple methods of detecting and quantifying some important pollutants in the terrestrial, atmospheric and aquatic environments. It is helpful to recognize two broad categories of approaches to monitoring pollution.

14.2 Monitoring with scientific apparatus

The monitoring of the actual quantities of a pollutant with scientific apparatus has the great advantage that it provides an indication of the actual quantities of substances involved and (in theory) facilitates rapid and reliable comparisons from place to place and time to time.

However, several difficulties arise. It is not always simple to identify which substances are the cause of a particular problem. Physical and chemical analyses may become expensive, requiring properly staffed and equipped laboratories. Difficult decisions have to be made on the location of monitoring or sampling sites and the frequency with which samples are taken. There may be major problems in establishing the representativeness of the sample and sampled locations and in determining the hour by hour, day by day or seasonal fluctuations in pollutant type and concentration. The cost in salaries and analytical procedures adopted may be difficult to sustain. In addition, many environmental monitoring procedures exist which can generate data at such an alarming rate that substantial further investments in data handling and analysis are needed.

14.3 Biological indicators

The survey and mapping of biological indicators as a pollution study method follow directly from the fundamental ecological principles concerning limiting factors and productivity described in Chapters 2 and 6. This approach concentrates on mapping variations in species types, species diversity and population numbers. Such surveys may map the distribution of these biological properties within and around pollution sources, or establish how these properties compare, before and after the establishment of a pollutant source. The advantages of this approach are many. It is relatively quick, inexpensive and requires relatively few people compared to laboratory studies. It investigates the actual biological situation in the field and because the species will be affected by short-term as well as medium-term variations in pollution concentrations, the problems of establishing the representativeness of samples taken at specific periods are to some extent by-passed.

However several other issues must be resolved in pollution surveys. The most important are to determine whether the biological variations found are a response to the pollutant or other factors; to clarify the related question of the nature of the 'baseline', i.e. 'natural' habitat conditions in the area (an important reason for the conservation of pristine, natural, unpolluted ecosystems). It is necessary to be able to identify the type and status of the biological indicator species. The extent to which the pollution source is the major cause of the variation must be established and whether or not the source may be tipping the balance in a situation caused by longer-term changes, which result from other human activities or natural environmental change.

The more reliable biological indicator species may not occur naturally within the study region. Consequently there may be problems of comparison or extrapolation from one area to another.

Finally, at some point in the study, it will be necessary to calibrate the biological changes detected with actual measurements of the amounts of the pollutant involved, in order to be able to comment reliably on the broader significance of any problems detected.

14.4   Pollution of the terrestrial environment

Soils

Soils may become polluted as a result of the deposition of material from the atmosphere or from the addition of agricultural chemicals. Soils prone to flooding may be affected by substances deposited from polluted water. In some ways, agricultural chemicals are the most insidious because they are applied for a 'good' reason; commonly they are very complex substances which require specialized equipment and skills to detect.

Spoil tips and derelict land

Very polluted soils can exist in these locations which are often a conspicuous legacy of both past and present mining and industrial activity. Two types of spoil tips can be recognised: metalliferous and non-metalliferous.

Metalliferous spoil tips occur in many parts of the world. In the United Kingdom they are well represented in the Tamar Valley and Devon and Cornwall generally, Derbyshire, Cumbria and South Wales. The gaunt pumping houses and quarried floors which are attractive to so many tourists in the South-West are often associated with soils polluted by a wide variety of ions such as lead, arsenic, cadmium and zinc. The soils are therefore toxic to many species of plants, a problem which is often compounded by a lack of essential macro- and micro-nutrients. However, special varieties of plant species or ecotypes are often found to be adapted to these difficult substrates. The accumulation of toxic cations through the food webs developed within and upon these soils may be very important.

Non-metalliferous spoil includes china clay waste and colliery wastes, household refuse and many municipal waste tips. In general these spoil tips also lack vital nutrients. The diversity of species is limited by such factors and ecosystems of low productivity are produced.

In both cases a major problem is to prevent the run off of polluted water. This is particularly important where heavy metals such as lead or cadmium are washed into other habitats, and around colliery spoil where the breakdown of pyrites leading to acid products may cause very acid run off (Section 6.4).

The reclamation of these tips presents many problems. Some may be retained as ugly, but challenging areas for trail bikes and other motorized recreation. Otherwise, the major objective is to facilitate the growth over the spoil of a continuous layer of vegetation, perhaps grass pasture, in order to improve the appearance of the environment, to reduce problems of leaching, and to prevent erosion by wind or water transferring harmful substances into the surrounding area. An important approach is to change the shape of the tip so that it is less prone to soil erosion (a drumlin appears to be the most favoured shape) and then to cloak the polluted soil with imported topsoil. A seed mixture may then be applied in a mulch or nutrient-enriched medium to provide a rapid and fairly continuous grass cover. These procedures are expensive. In the past, problems of die-back have occurred as a result of well-established herbs or trees eventually extending their roots down to the toxic levels.

A less expensive approach is to try to find out which species of plants will grow satisfactorily on the polluted ground. Hopefully the species will be able to sustain themselves for sufficient time so that, on one hand, some toxic elements will be lost through natural leaching, while on the other, the slow accumulation of organic debris will start to provide a trap and reservoir for both nutrients and soil moisture – properties lacking in most highly porous, drought-prone, nutrient-deficient toxic spoils. In this context, the bioassay methods described in Chapter 6 provide both a measure of the toxicity to the plant of the pollutant(s) in the soil and a method for identifying which are the better species with which to begin reclamation trials.

Lead emission from motor vehicles

The pattern of increase in lead concentration in the atmosphere measured by entrapment in the upper layers of the Greenland ice sheet during the twentieth century is illustrated in Figure 14.1. This is largely attributable to the combustion of leaded fuels used in motor vehicles. Lead occurs naturally in soil, air, water and organisms. The crust of the Earth contains an average lead concentration of 20 mg/g which reaches the soil by rock weathering and human activities. Thornton (1980) estimated that approximately 4000 square kilometres of agricultural land in England and Wales were badly affected as a result of past mining and smelting. The Royal Commission on Environmental Pollution (1983) surveyed the soils in 10 British towns and found an average soil lead concentration of 671 mg/g. Rossin *et al.* (1983) noted that more than 2000 tonnes of lead reach sewage plants, from domestic and industrial sources and surface run-off in the UK. Eventually much water-borne lead reaches the

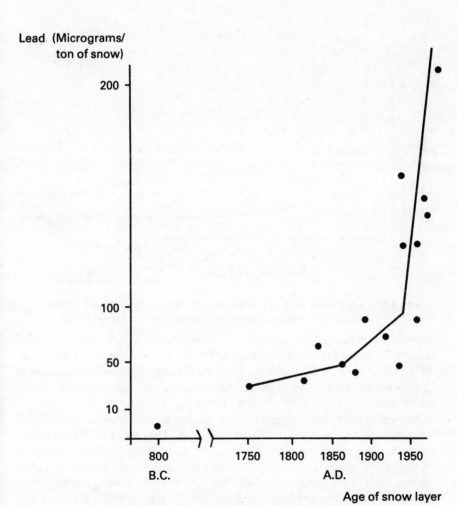

Figure 14.1 Lead content of snow cores from the Greenland ice sheet (Redrawn with permission from Patterson and Salvia (1968).)

sea. Despite this, Norton (1982) showed that most of the lead in the North Sea arrived as a result of atmospheric deposition (4500 tonnes annually) from nearby countries: 2400 tonnes is calculated to be introduced by rivers.

With the exception of aquatic systems, lead is seldom found in the ionic form. Alkyl lead compounds are found in petrol and these may be absorbed by mucous membranes and the skin. The main pathways into organisms are through the respiratory and gastro-intestinal tracts. Various calculations have indicated that urban adults receive approximately 30 micrograms of lead daily through the gastro-intestinal tract and about 20 milligrams through the respiratory system. As the lead intake increases, so too does the amount of lead retained in the body. Lead retention is low, but with prolonged exposure,

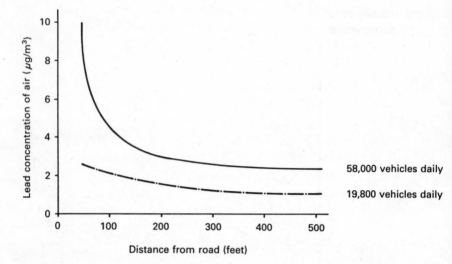

Figure 14.2   Lead concentrations in relation to distance from a road (Redrawn with kind permission of the American Chemical Society from Daines *et al.* (1970).)

clinical systems of lead poisoning will eventually appear. The lead will accumulate in a variety of organs, but 80–90 per cent will be found in the skeleton. Clinical lead poisoning is characterized by headaches, fatigue, loss of appetite, anaemia and severe abdominal cramps. Lead affects both haemoglobin synthesis and red blood cell survival times in the body.

Several workers have attempted to investigate the pattern of lead dispersal in the environment from motor vehicle emissions. Takala and Olkkonen (1981) working in an urban area in central Finland indicated that the lead content of the lichen *Hypogymnia physodes* was related to motor vehicle density and to distance from roads. Studies by Daines *et al.* (1970) also showed this pattern of decrease with increasing distance from source (Figure 14.2). Harrison and Laxen (1978) demonstrated the extent of atmospheric pollution by lead over rural areas that could be traced to urban sources.

The detection of lead in plant tissues

Glater and Hernandez (1972) have described a simple and rapid method for the identification of lead in plant tissues. It may therefore be used to trace part of the lead pollution pathway in the biosphere. The technique employs the chemical sodium rhodizonate ($C_6O_6Na_2$) which forms a scarlet precipitate with lead ions in an aqueous solution buffered at pH 2.8 ($2PbC_6O_6$ $Pb(OH)_2H_2O$).

Equipment
Biology/Chemistry laboratory equipped with a fume cupboard; deionized

water, sodium rhodizonate, tartaric acid, sodium bitartrate, pH meter, pH buffers, dropper bottle, wash bottles, burette, glass beakers, scientific balance, stirring rods, glass microscope slides, cover slips, light transmission microscope ($\times$100–400 magnification).

Procedure
The procedure must only be carried out under the direction and supervision of a competent chemist/biologist who must be familiar with the procedure and the necessary safety and emergency procedures in the event of an accident.

1   The sodium rhodizonate reagent should be freshly prepared by dissolving 50 mg of sodium rhodizonate in 25 ml of deionized water.
2   A buffer solution is prepared by dissolving 1.5 g of tartaric acid and 1.9 g of sodium bitartrate in 100 ml of deionized water.
3   Using a very sharp blade or scalpel, obtain thin slices or squashes of the plant tissue. Place them in approximately 2 ml of sodium rhodizonate solution and leave for 30 minutes. Add a few drops of buffer solution, swirl the mixture and leave for a further 10 minutes.
4   Rinse the material with deionized water, mount in the buffer solution on a glass slide and examine immediately with the microscope.

When lead is present in the tissue an intense, bright pink to deep scarlet colour results. This examination must be carried out immediately because the colour fades on standing.

14.5   Pollution of the atmospheric environment

Pollutants escape from chimneys in the various ways illustrated in Figure 14.3. A variety of factors influence the discharge and subsequent dispersal of the pollutants, these include:

1   topography;
2   the incidence of climatic inversions – these tend to trap pollutants close to ground level;
3   the concentration of pollutants in the atmosphere and the various concentrations likely to prove toxic to organisms;
4   wind velocity which can cause dilution, together with direction which interacts with local topography to influence the pattern of dispersal;
5   precipitation, both the amount and type;
6   height of the discharge of the primary pollutants from the stack;
7   the projected plume impact point and its variation with atmospheric conditions;
8   reactivity – primary pollutants may react in the atmosphere to form secondary pollutants, some of which may be more dangerous than the primary pollutants;

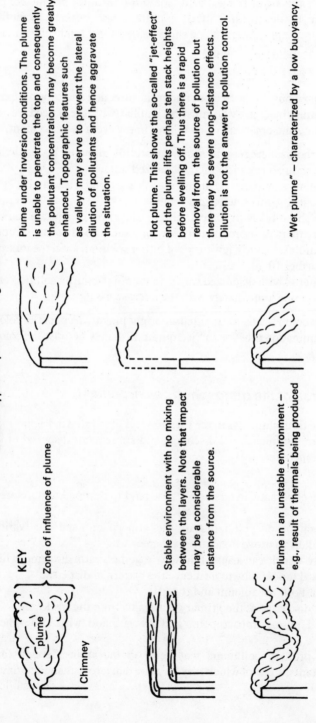

General types of plume – diagrammatic

KEY

Zone of influence of plume

plume

Chimney

Plume under inversion conditions. The plume is unable to penetrate the top and consequently the pollutant concentrations may become greatly enhanced. Topographic features such as valleys may serve to prevent the lateral dilution of pollutants and hence aggravate the situation.

Hot plume. This shows the so-called "jet-effect" and the plume lifts perhaps ten stack heights before levelling off. Thus there is a rapid removal from the source of pollution but there may be severe long-distance effects. Dilution is not the answer to pollution control.

"Wet plume" – characterized by a low buoyancy.

Stable environment with no mixing between the layers. Note that impact may be a considerable distance from the source.

Plume in an unstable environment – e.g., result of thermals being produced

Figure 14.3   A diagrammatic representation of different types of smoke plume

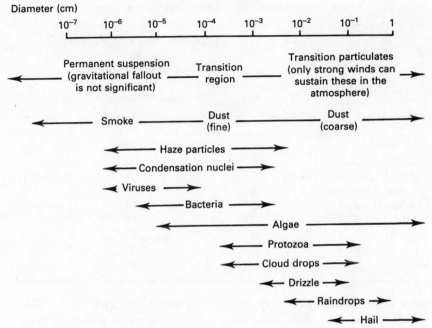

Figure 14.4 The relative diameters and settling velocities of particles. As the diameters of the suspended particles increase, the rate of fall increases and gravitational settlement becomes more important (After Patterson (1973) and redrawn with kind permission of W. H. Freeman and Co., from Moran, J. M., Morgan, M. D. and Wiersma, J. H., *Introduction to Environmental Science* © 1980. All rights reserved.)

9 temperature of the gases at discharge: this influences the height they will reach in the atmosphere and hence dispersal patterns;
10 the susceptibility of individuals, species and ecosystems to pollutants;
11 the available treatments, pollution monitoring and abatement measures.

Particulate matter

Particulates are produced in various forms – ash, soot, dust and smoke. They result from a wide variety of industrial and domestic sources as well as from the erosion of soil and spoil tips. The fate of particulate matter in the atmosphere is affected by prevailing climatic conditions and particle size (Figure 14.4). The larger particles generally sediment or impact fairly close to their source.

In recent years there has been a trend towards a decrease in the quantity of smoke pollution in the UK. This is usually attributed to more efficient fuel combustion and efficient smoke arresters. Tables 14.1 and 14.2 provide data on the relative importance of smoke density and undissolved deposition at

Table 14.1   *Urban air pollution by smoke* (After Warner *et al.* (1969).)

| | Average daily smoke density concentration ($\mu g/m^3$) Year ending March | | | |
| | 1966 | | 1967 | |
Locality	winter	summer	winter	summer
Aylesbury, Town Hall	80	25	91	24
Bath, Fire Station	63	23	53	27
Cardiff, City Centre	37	20	38	23
Harrogate, Municipal Offices	106	33	75	36
London, Kensington	95	39	73	44
London, County Hall	95	39	74	58
London City: St Pauls	77	33	65	37
Sheffield	104	49	82	44
Ebbw Vale, Site 4	77	39	74	36
Port Talbot, Site 9	29	21	29	17

Table 14.2   *Urban air pollution by undissolved deposition* (After Warner *et al.* (1969).)

| | Undissolved deposition, $tons/mile^2/month$ Year ending March | | | |
| | 1966 | | 1967 | |
Locality	winter	summer	winter	summer
Cardiff, City Centre	8.86	4.54	7.55	6.78
Harrogate, Municipal Offices	2.46	4.47	5.00	3.62
London, Guildhall	24.18	18.63	21.10	15.40
Ebbw Vale	55.76	40.35	30.28	40.35
Port Talbot	39.01	37.56	26.62	33.16

various sites in the British Isles. There is clear evidence of a decrease in pollution concentrations occurring during the summer.

Particulates around a steel works
More detailed information on fluctuations in the concentration of smoke at seven sites in the steel producing town of Port Talbot in South Wales is presented for comparison in Figure 14.5 (a) and Tables 14.3 and 14.4. These data emphasize:

1   the importance of horizontal transport of particulates being blown and caught by plants;
2   the differences in pollutant catch that exist between species as a result of the shape, size and presentation of their leaves;

Table 14.3 *Leaf contamination by particulate pollutants* (After Pyatt (1973).)

Material	Site	Weight of particulates as % of air dry weight		
		on entire leaf	on upper surface	on lower surface
Laurel*	1	1.92	1.31	0.61
Ribes*	1	4.84	2.60	2.24
Lonicera*	1	1.71	1.60	0.11
Grass	1	0.30	0.13	0.17
Snowberry*	1	1.15	1.09	0.06
Ground ivy	1	1.37	1.02	0.35
Buddlea*	2	0.37	0.25	0.12
Berberis*	2	4.13	3.50	0.63
Laurel*	2	0.50	0.40	0.10
Hydrangea*	2	0.51	0.34	0.17
Wallflower	2	1.31	0.79	0.52
Beech*	3	0.12	0.07	0.05
Peach*	3	0.73	0.41	0.32
Grass	3	0.22	0.10	0.12
Oak	4	0.39	0.23	0.16
Hawthorn*	4	0.73	0.51	0.22
Blackberry*	4	0.85	0.64	0.21
Hazel*	4	1.10	0.86	0.24
Blackberry	5	0.71	0.59	0.12
Oak	5	0.66	0.45	0.21
Hawthorn*	5	0.56	0.42	0.14
Alder*	5	0.72	0.52	0.20
Blackberry*	6	0.61	0.57	0.04
Alder*	6	0.42	0.29	0.13
Oak	6	0.43	0.29	0.14
Hawthorn*	6	0.19	0.12	0.07

Notes: Material marked * was sampled at a height of 4 ft. The other material, with the exception of grass, wallflower and ground ivy, was collected at a height of 7ft. (See Figure 14.5(b).)

3 that transport of particulates occurred well beyond the area of the steel works.

Effects of airborne particulates

Particulate matter is an important pollutant in both urban and rural areas. Particulates have a variety of effects. For example, they adversely affect visibility and will stain a variety of substrates and fabrics. They also have important biological effects. They can block stomata on plant surfaces; they can 'smother' established species and occasionally may create an unstable or toxic substratum. For example, the net assimilation rate of laurel leaves from polluted sites has been shown to be much reduced compared to those from

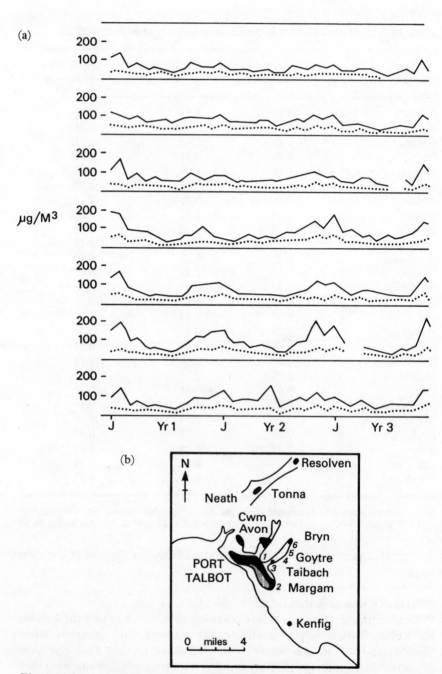

Figure 14.5(a)   Fluctuations in the concentration of smoke at sites in Port Talbot, South Wales, over a period of three years J = January; continuous line = highest daily value; broken line = mean value (After Pyatt (1973); with kind permission of Gordon and Breach Publishers Ltd.); *14.5(b)* The port Talbot area.

Table 14.4   *Effect of leaf position on accumulation of particulates* (After Pyatt (1973).)

Species	Site	Height (ft)	Weight of particulates as % of air dry weight on entire leaf	on upper surface	on lower surface
Laurel	1	3	2.26	1.69	0.57
		6	1.83	1.29	0.54
Hydrangea	2	3	0.63	0.49	0.14
		6	0.50	0.35	0.15
Oak	5	5	0.63	0.42	0.21
		12	0.51	0.35	0.16
Oak	6	5	0.42	0.29	0.13
		12	0.32	0.21	0.11

non-polluted areas. In man, the mucous epithelium traps much of the particulate matter and smaller particles can penetrate deep into the lung itself. Some species trap more particulates than others. Evergreen trees can be more efficient 'pollution' traps than deciduous trees in winter and in very polluted areas they may be killed by high concentrations of particulates. Some species adopt a deciduous growth strategy in response to the problem (Pyatt 1973).

Heavy metals may be transported with the particulates. In industrial Severnside, Little and Martin (1972) found that concentrations of these metals were consistently higher in elm leaves rather than hawthorn leaves. Concentrations were also found to be as great, if not greater, on the sheltered side of the tree. Whereas most lead introduced in this way could be washed off the leaf, only 45 per cent of the zinc and 28 per cent of the total cadmium could be similarly removed, even with the aid of a strong detergent.

Determination of the amount of particulate matter in the atmosphere

The best method is to utilize a British Standard deposit gauge, a 'Warren Spring Apparatus' or its overseas equivalent. However, inexpensive and simple equipment can be assembled and used at various distances and directions from the pollution source.

Adhesive strips and cards

The simplest method is to expose adhesive strips to the atmosphere. These should be protected from the direct impact of rain to prevent the washing of particulates from the tape. The tape is then examined under the microscope and an estimate of the size and number of particles made with the aid of a graticule eyepiece. Usually, the particles with the largest diameter settle out nearest the source. The 'zone of influence' of various point sources under different environmental conditions can then be determined and affected areas mapped.

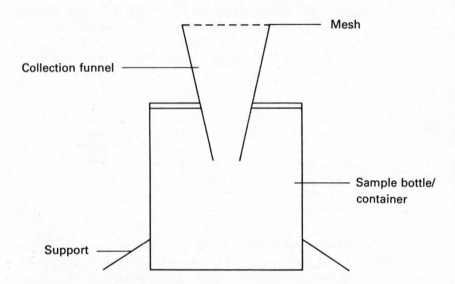

Figure 14.6   A sampler for particulate matter

Gridded, adhesive cards specifically designed for dust particle collection are manufactured by the Esselte Studium Company and may be purchased from biological suppliers (see Appendix).

Sampling vessels
A home-made particulate sampling vessel is illustrated in Figure 14.6. In its simplest form the apparatus comprises a graduated sampling bottle which stores the particulates and precipitation captured by a funnel. A 5 mm mesh will prevent leaves and other detritus entering the funnel. The vessel may be supported by bricks or suspended in a wire cradle.

In the laboratory, the contained sample is agitated to loosen particulates adhering to the sides of the chamber. The volume of precipitation captured should be recorded. If necessary, wash material off the sides of the bottle with distilled water from a wash bottle. Transfer the water and suspended particulates to a beaker of known weight and evaporate to dryness. The weight of solids caught is obtained by re-weighing the beaker with the dried solids and subtracting the initial weight of the beaker on a scientific balance accurate to 0.01 g or 0.001 g. The results are expressed in terms of mg of particulates. To express particulates per unit area, the diameter of the funnels must be measured. The catch area is the surface area of the mouth of the funnel.

The Warren Spring apparatus
This device is illustrated in Figure 14.7. It is extremely useful for detailed studies of air pollution by particulates. Air is drawn through a clamp which

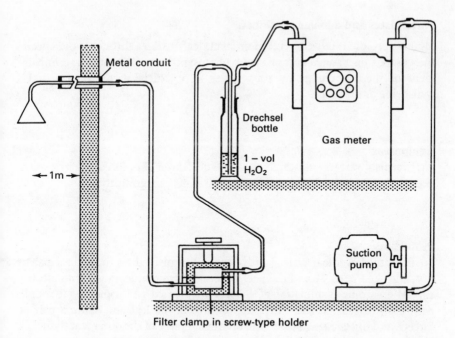

Figure 14.7 The Warren Spring apparatus – schematic arrangement of the standard daily smoke or sulphur dioxide sampling apparatus (Redrawn with kind permission of Dept of Trade and Industry, Crown Copyright reserved.)

contains a filter paper. The air then passes into a Drechsel bottle with hydrogen peroxide. The actual volume of air sampled in a 24 hour period is recorded automatically on a meter.

Place a clean filter paper on a white tile background. Test with a reflectometer and adjust so that its meter reads 100. (The supplier of the EEL reflectometer is given in the Appendix.) Replace the clean filter paper with one that has been exposed to particulate matter in the Warren Spring apparatus for 24 hours. Consult the conversion tables supplied with the reflectometer and these will provide a scale with which to convert and calibrate the reading obtained in terms of particulate pollution. Alternatively, your local Environmental Health Officer may be able to supply a conversion scale.

Sampling strategy

The decision on where to locate the sampling vessels in the field must recognize that the gauges need to be a standard height above ground, for ideal sampling, to avoid catching soil splash, animal additives and still remain accessible. Trees and buildings generate eddies and turbulence which can produce unrepresentative results. The safety of the sampling equipment will be of much importance in the selection of sampling sites; other include considerations of pollution dispersal, climatic conditions and the local topography.

Particulates and damage to stomata

Stomata are the minute openings on plant leaf surfaces through which occur the gaseous exchanges necessary for photosynthesis. Obviously the impairment of these openings by particulates is of crucial significance for the well-being of the plant.

Equipment
Nail varnish, razor blades, microscope slides, cover slip, forceps or tweezers, petri dish, transmission microscope ($\times 100$–$\times 400$) magnification.

Procedure
1   Paint the lower leaf surface with clear nail varnish, allow the nail varnish to dry, and then peel. Mount the peel on a microscope slide, cover with a cover slip. Microscopic examination of the peel will give a useful, general, idea of the nature of the leaf cuticle. Alternatively, break the leaf and using a pair of forceps and/or the razor blade, pull off a small piece of the outer leaf tissue.
2   Prevent the material from drying by mounting the specimen on the microscope slide in water.
3   Examine under the microscope.

The results from this simple test may be very interesting. For example, examination of the leaves of laurel from site 1, near the steel works in Figure 14.5(b), revealed that no less than 50 per cent of the stomata investigated were contaminated by particulates.

Gaseous pollutants: background

Sulphur dioxide is a well studied atmospheric pollutant. Estimates of worldwide sulphur dioxide emissions put the atmospheric contribution from fuel burning as responsible for 16 per cent of the total atmospheric SO_2. The actual concentration of sulphur dioxide in the atmosphere is likely to vary significantly from place to place and time to time. For example, studies concerned with the occupational health of blast furnace workers in South Wales recorded maximum concentrations of 13,129 micrograms per cubic metre near the blast furnace (Lowe and Campbell, 1968). The polluted air in the adjacent town of Port Talbot gave values of between 34 and 85 micrograms per cubic metre. Pollutant concentrations were less in the summer months. Onshore winds caused a rapid dilution of the pollutants at this time.

The efficiency of the combustion of fossil fuels can also alter the quantity of SO_2 liberation. When coal is burnt, carbon dioxide, water, sulphur dioxide, oxides of nitrogen and mineral matter are important products. If combustion is relatively inefficient, then carbon monoxide and even elemental carbon may be produced. More efficient combustion is being encouraged at present and is likely to lead to the enhanced release of oxides of sulphur and of nitrogen. These will need removal by 'scrubbers'.

The nature of the problem is likely to intensify in the future. At present, good quality fuels are burnt. These have a natural sulphur content of between 1 and 3 per cent, although a number can contain sulphur in excess of 10 per cent. However, in view of current energy problems and the likely future demand from industrializing countries it is probable that poorer grade, higher sulphur fuels will be more widely used in this country.

Effects of sulphur dioxide on plants
Middleton, Emik and Taylor (1965) give four criteria to describe the effects of pollution on biological material. These are:
1 interference with enzyme systems;
2 change in cellular chemical constituents and physical structure;
3 retardation of growth and reduction in production from altered metabolism;
4 acute and immediate tissue degeneration.

Damage to leaves
The marginal and interveinal areas of leaves are susceptible to sulphur dioxide pollution, generally becoming dull in appearance and then white or reddish-brown. The surrounding tissue remains green.

If the exposure is prolonged, or if a high concentration of sulphur dioxide is present, then mesophyll cell plasmolysis – loss of water leading to cell content contraction – and injury to the photosynthetic palisade and spongy mesophyll cells may occur. Leaves may be prematurely senescent and shed. This is especially true of the needles of coniferous trees.

Factors influencing the impact of sulphur dioxide on plants
The following factors, acting singly or in combination, are of importance in this context.

1 Concentration and exposure duration. Each species has a threshold concentration, usually designated C_R which must be exceeded before damage occurs. In the field, the plant will be exposed to fluctuating concentrations of the pollutant. A high concentration for a short period tends to have a greater effect than a lower concentration for a correspondingly longer period. It has been

shown that the periods when concentrations are below the irritation threshold concentration, may act as recovery periods for the plant.

2   Species. Thomas and Hendricks (1956) and others, have noted that different species have different sensitivities to sulphur dioxide. They categorize species according to their sensitivities:

(a) 'sensitive' species include lucerne, lettuce and oats;
(b) 'intermediate' species include tomato, cabbage and peas;
(c) 'resistant' species include citrus and privet.

3   Stage of growth. Thomas and Hill (1935) working with lucerne noted that the loss in yield in dry matter was proportional to the amount of leaf area destroyed. However, other workers have recorded that the susceptibility and yield reduction were affected by the stage of growth of the plant when exposed.

4   Time of day. Thomas (1958, 1961) found the susceptibility of plants to be greatest from mid-morning to mid-afternoon. Most plants were not susceptible at night when stomata are closed.

5   Environmental considerations. Plant species have been observed to be more susceptible to sulphur dioxide when environmental conditions favour stomatal opening. Consequently moisture stress tends to increase the resistance of plants to sulphur dioxide (Oertli 1956).

Measurement of sulphur dioxide concentration in the atmosphere

1   A simple method of monitoring sulphur dioxide concentration in the atmosphere is to use the standard lead dioxide candle. This candle integrates, over a period, the reaction which occurs between sulphur compounds in the air and a surface of lead dioxide prepared in a specific way. The rate of reaction is affected by various factors in addition to sulphur dioxide values: air temperature, atmospheric humidity, wind speed. Consequently, for comparative studies, it is necessary to ensure the measurements are taken in standardized conditions. Suppliers of lead dioxide candles are given in the Appendix.

2   A Dräger pump and tubes is illustrated in Figure 14.8 (see Appendix for suppliers) and is used in connection with a simple bellows pump. The pump is hand operated and sucks in exactly 100 cubic cms of gas at each stroke. It requires practically no maintenance. The Dräger tube is a glass tube which is fused at both ends and contains appropriate chemicals for gas determination. For measurement, both tips of the Dräger tube are broken off and one end of the tube is inserted into the pump. The prescribed number of strokes are made sucking air through the tube. The colour generated by the chemical change within the tube may then be evaluated using the graduations on the tube itself.

3   The Warren Spring apparatus (Figure 14.7) can be used to monitor air pollution as well as the particulate content of the atmosphere. Air which has

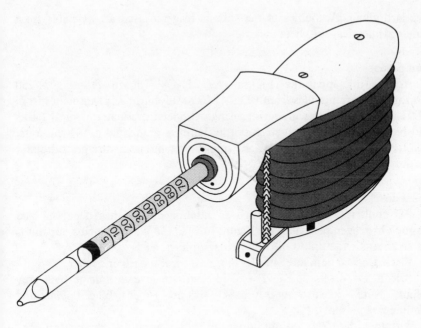

Dräger Tube

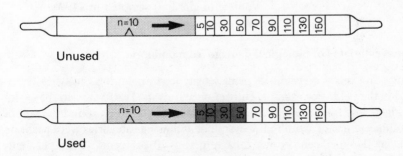

Unused

Used

Figure 14.8   The Dräger pump and tubes. (Redrawn by kind permission of Draeger Safety (1978).)

been filtered by being pumped through a filter paper is then drawn into the Drechsel bottle which contains hydrogen peroxide solution, which serves to remove the sulphur dioxide. The samples are best collected every 24 hours and the volume of air sampled is read from the reading on the gas meter. The amount of sulphur dioxide generating acidity is determined by titration with a standard alkali. Ampoules of the various reagents used are available from standard chemical suppliers.

Equipment

Warren Spring apparatus (Figure 14.7); H_2O_2 – 1 vol. strengh; N/250 sodium borate (ampoules from BDH); N/250 sulphuric acid (ampoules from BDH); distilled water; glass measuring cylinders; graduated conical flasks and beakers; small funnel; wash bottle; burettes graduated in 0.1 ml units; burette stands; white tile; BDH '4.5' indicator solution in dark glass dropper bottles.

Procedure

1 Air containing pollutants such as sulphur dioxide has been bubbled through hydrogen peroxide (see Figure 14.7) in the Warren Spring apparatus for 24 hours. The sulphur dioxide yields sulphuric acid.

2 Place exposed peroxide samples in a glass conical flask or beaker.

3 Add a few drops of BDH '4.5' indicator to give a pink or pink-grey colour. (Note – it will stay grey if SO_2 was not present; blue if something alkaline was present).

4 Titrate with N/250 sodium borate (from the burette) to the neutral–grey end point.

5 The SO_2 concentration ($\mu g/m^3$) is then calculated:

$$SO_2 \text{ concentration} = \frac{4520 \times [n\% \text{ of ml. N/250 sodium borate added}]}{\text{Volume of air (Ft}^3\text{) sampled in 24 hrs.}}$$

Visual estimates of biological damage from pollution

Rough-and-ready divisions of plant sensitivity to pollution can be tested or used in the field. For example Gilbertson and Pyatt (1980) mapped the extent of foliage loss from the crowns of trees and shrubs, in and around the area of Whitehaven on the Cumbrian coast. The following categories were used: no damage, 1–5 per cent crown foliage loss; 5–10 per cent; 10–25 per cent; 25–50 per cent; 50+ per cent. The results are illustrated in Figure 14.9.

These data suggest that both the pattern and intensity of crown foliage loss displayed by certain tree species might be the result of the interaction of strong, salt-laden, onshore winds with atmospheric pollution from industrial sites.

The pattern of damage suggested that the local wind system was strongly influenced by topography. It was also possible to rank the extent of canopy loss in various tree genera by simple observation (Table 14.5). Clearly the planting of trees for amenity purposes in the area would have been more successful if more attention had been paid to the different tolerances of tree

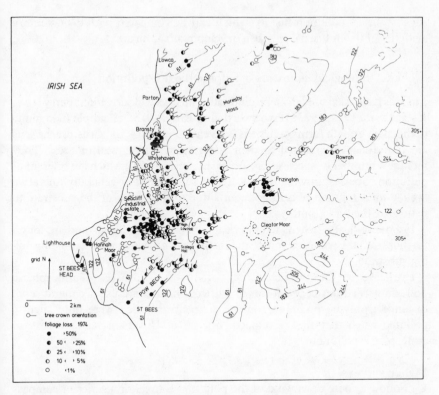

Figure 14.9 Tree and crown foliage loss and tree crown orientation on the coast of Cumbria, North-West England (Gilbertson and Pyatt (1980); with kind permission of the Journal of Biological Education.)

Table 14.5 *Canopy loss in various tree genera around the coast of Cumbria* (After Gilbertson and Pyatt (1980).)

Greatest canopy destruction	*Least canopy destruction*
Pinus spp. *Betula* spp. *Acer pseudoplatanus* *Larix* spp.	*Quercus* spp. *Ulmus* spp. *Crataegus monogyna* *Salix* spp. *Fraxinus excelsior*

and shrub species and to the importance of shelter from salt-laden or polluted winds.

Growth increment studies

Pollution damage may affect the productivity of the tree or shrub. This is likely to be evidenced in the growth increment history of the tree. Studies of tree growth variations over time following the methods described in Chapter 6 may suggest an air pollution history which can be compared with documentary evidence of the industrial or urban developments in an area.

Lichens – biological indicators of sulphur dioxide pollution

Lichens have been used in many air quality surveys (Gilbert, 1965; Ferry *et al.*, 1973; Pyatt, 1970, 1973; Hawksworth and Rose, 1976). A simple field guide to the main growth forms of lichens is given in Figure 14.10. 'Crustose' lichens are found closely attached to their substratum – tree, wall or rock. They include a number of species (e.g. *Lecanora conizaeoides*), which are tolerant of pollution. Species growing on calcareous substrates can generally tolerate a higher concentration of acid pollutants due to the ability of the substrate to neutralize the acid input.

Foliose or 'leafy' lichens tend to be less tolerant of pollution and are found further from the pollution source unless they are growing on an acid-neutralizing substrate.

Fruticose or 'bushy' lichens are often very sensitive to atmospheric pollution. Well-developed forms of fruticose lichens are usually found some distance from the pollution source, or in areas which were only slightly polluted when they first developed. Species of *Usnea* are very sensitive to atmospheric pollution.

Lichen surveys

1   Follow a route downwind of the pollution source, or a series of compass bearings which radiate from the pollution source; and inspect the lichen flora at a fixed height, substrate, and orientation, as convenient, certainly every 0·5 km.

2   For each principal substrate at each sample site note:

(a) the morphological forms present i.e. crustose, foliose and fruticose and their relative cover-abundance;

(b) the size and cover value of lichen thalli using a 10 cm × 10 cm gridded transparent plastic sheet;

(c) if possible identify the lichen species present using Alvin and Kershaw's book (1963).

14.6   Pollution of freshwater environments

Eutrophication is the enrichment of waters by nutrients brought in from a variety of sources. The conversion from a nutrient-poor (oligotrophic) water body to a nutrient rich (eutrophic) one is generally a prolonged process, but is hastened by pollution from sewage and agricultural fertilizers. Eutrophication is a major problem of the freshwater habitats of the British Isles which have been described by Smith and Lyle (1979). The nutrient enrichment causes an

ZONE	0	1	2	3	4	5	6
LICHEN FORM	No Lichens	Crustose	Foliose	Foliose	Foliose	Fructicose	Fructicose
INDICATOR SPECIES	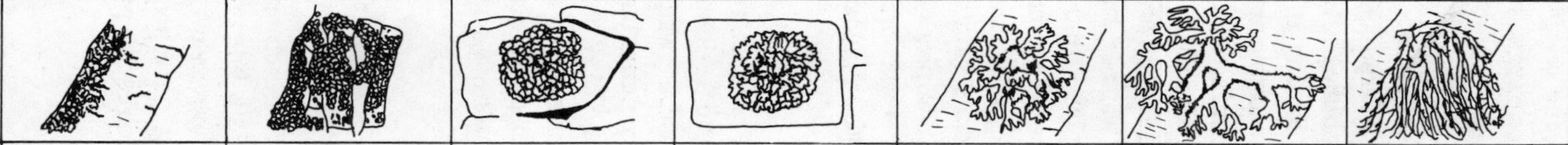						
DESCRIPTION	*Pleurococcus* Green powdery alga on trees	*Lecanora conizaeoides* Grey–green with a crazy paving appearance on trees and acid stone	*Xanthoria parietina* Bright orange on alkaline substrates for example, limestone, mortar and asbestos	*Parmelia saxatilis* Pale grey–green on old walls of an acidic nature	*Parmelia (Hypogymnia) physodes* Grey coloured On trees with other leafy lichens	*Evernia prunastri* Upper surface grey–green with a whitish lower surface, on trees where the air is almost free from pollution	*Usnea* sp. Beard lichens grey–green filamentous thallus, abundant in clean air

Figure 14.10 A simple scale of lichen growth forms and species used to indicate concentrations of atmospheric pollution (Adapted from the Advisory Centre for Education zone scale.)

explosion in the number of organisms in the water, notably initally in the population of freshwater algae. The need for oxygen for respiration and to aid the microbial decomposition of dead organisms may outstrip the rate at which it is added to the water by aquatic plants or incorporation from the atmosphere. In these circumstances the water may temporarily incur a low dissolved oxygen concentration – there is an oxygen debt. The introduction of biological remains to the water which require significant microbiological activity to cause them to break down may have the same effect of causing an oxygen deficiency. An indication of the potential of biologically oxidizable matter to de-oxygenate the water is the BOD (Biological Oxygen Demand).

There are various other environmental controls on BOD which, if altered, may have an adverse or beneficial effect on the aquatic system. For example, cool water will hold more oxygen than warmer water. Consequently the introduction of warm water from cities or power stations into streams may be considered as a further cause of pollution. The introduction of non-inert biological (e.g. sewage) suspended particles may cause oxygen depletion, as well as cause other changes in the stream's flora and fauna which may culminate in the so-called 'sewage-fungus complex'. Inert suspended solids, such as china clay or colliery tip waste, may settle out in areas of slow flow and smother established species, as well as create an unstable substrate unsuitable for many species. These wastes may also reduce the light available for photosynthesis and hence productivity.

Oils can form a surface film which may reduce re-oxygenation from the atmospheric reservoir. Detergents may cause unsightly frothing and dissolve the attachment discs or structures of various freshwater organisms.

The impact of toxins, acids, alkalis and the numerous other chemicals which enter streams, rivers and lakes are not always easy to predict. The toxicity of the substance and the capacity of the water body to dilute the substance are important. Some chemicals act preferentially on 'target species'. This loss may cause a temporary lack of competition for resources and a population explosion or 'bloom' in other remaining species.

Primary production in aquatic ecosystems

Light energy is used by the pigments of the chloroplast to form organic compounds by photosynthesis. This process is known as primary production (Section 6.1). An estimate of primary productivity may be made by measuring the amount of oxygen production occurring in freshwater ecosystems. This is obtained with a dissolved oxygen meter which can be purchased from scientific suppliers (see Appendix).

Equipment
500 ml clear glass, sealable bottles; aluminium foil; dissolved oxygen meter.

Procedure
1   Collect two 500 ml samples of water in a glass bottle.
2   Determine the dissolved oxygen content of both samples using the meter.
3   Seal both sample bottles.
4   Wrap one bottle in aluminium foil to exclude light completely from the contents of the bottle.
5   Place both bottles at the collecting site and leave for a minimum period of 2–3 hours.
6   Determine the dissolved oxygen content of both samples.

In the wrapped bottle only respiration will have occurred as light has been excluded; while in the unwrapped bottle both respiration and photosynthesis should have taken place. The total quantity of oxygen produced in the sampling period is given by the difference between the dissolved oxygen concentrations at the start of the experiment and that recorded at the finish in each sample. This quantity of oxygen is therefore an indirect measure of the difference in amount of primary productivity induced in the samples. The use of the technique allows comparisons to be made between 'pristine' sites and those experiencing eutrophication or pollution.

Toxicity in freshwater ecosystems

Alderdice (1967) distinguished two broad types of toxic effect:
1   acute toxicity where a dose of poison in a short period is generally lethal;
2   chronic toxicity where a relatively low dose over a prolonged period may be lethal or sub-lethal.

The terms used here have been defined in biological terms by Sprague (1969):
(a)   acute – coming speedily to a crisis;
(b)   chronic – continuing for a long time;
(c)   lethal – causing death, or sufficient to cause it, by direct action;
(d)   sub-lethal – below the level which directly causes death;
(e)   cumulative – brought about, or increased by successive additions.

The American Public Health Association (1976) have provided a number of other useful terms:

1   Maximum allowable toxicant concentration – the concentration of a toxic product which can exist in a receiving water without impairing the productivity or other uses of the water.
2   Effective concentration – this is employed when an effect apart from mortality is being investigated; for example, respiratory stress. The results are expressed in terms of the percentage of animals affected at a particular concentration – the duration of the exposure is also important.

3   Safe concentration – the maximum concentration of a toxic material which has no observable effect on a species after prolonged exposure through one or more generations.

4   Incipient lethal level – the concentration at which acute toxicity ceases: i.e. generally the concentration at which 50 per cent of the population can survive for an indefinite time.

A standard laboratory test for toxicity uses the freshwater crustacean *Daphnia* (the water flea), which is sensitive to aquatic toxins. If a chemical is highly toxic to *Daphnia* then it would not be approved by the Ministry of Agriculture (UK) for use in water or near water. A standard test is to determine the lethal concentration which will kill 50 per cent of the population of water fleas in a given time – this concentration is known as LC_{50}.

Biotic indices of water quality

Many biologically-derived indices have been developed to assess water quality. These indices express the tolerance or sensitivity of an organism to pollution and values are assigned on this basis. The sum of these values gives an index of pollution for the site. One of these is the 'Trent Biotic Index' which was developed by Woodiwiss (1964). The sensitivity of various species is scored and summed. On this scale an unpolluted site will yield the maximum score of 10. Heavily polluted sites devoid of macro-intertebrates will yield a score of 0.

Measurement of important environmental variables

The nature and significance of the pollutants depends upon a complex of other environmental factors. Some of the more important are listed below. The accurate and precise measurement of each of these are major subjects in their own right, nevertheless, some simple, 'robust' procedures are set out.

Procedures

1   Temperature – a thermometer is suspended on a weighted line and lowered under the water for 2 minutes at a standard depth.

2   pH – standard portable, direct reading pH meter (see Appendix).

3   Dissolved oxygen – dissolved oxygen meter (see Appendix).

4   Water velocity – time the movement of oranges over a 10 m–20 m reach of water. There is a good correlation between measurements obtained in this way and flow metering of the stream velocity.

5   Discharge – the discharge of a stream is the product of its average velocity and its cross-sectional area.

6   Conductivity determinations are related to the quantity of dissolved and suspended salts in the water body.

7 Suspended solids – filter the water through a filter paper and note the colour difference. Unfortunately not all suspended solids are grey or black. One method is to determine their weight in a beaker after evaporation of the water. An alternative is to lower a weighted white disc of 20 cm diameter (a Secchi disc), into the water and to note the depth at which it cannot be seen.

8 Substrates – introductory texts on soils and sediments describe the necessary field and laboratory techniques (Briggs 1977a, b).

9 Marginal and aquatic vegetation – follow the procedures described in Chapters 4 and 5.

14.7 Acidification: interaction of the terrestrial, atmospheric and freshwater environments

In each of the preceding sections, the effect of acid substances has been described. These are linked together in the concept of acidification. This process has aroused great concern in most industrialized nations and can arise from many reasons; e.g. acid runoff from colliery tips or the discharge of acid industrial wastes into rivers. At the moment interest focuses on the phenomenon of acid rain. Unpolluted rain has a pH of 5.65 which is the same as that of distilled water in equilibrium with carbon dioxide. The term 'acid rain' is applied to rain more acidic than this i.e. the pH is below 5.6. The most important causal agents are the oxides of sulphur and nitrogen dissolving in precipitation. The components of precipitation acidity have been determined by a number of research workers and typical estimates of its components are sulphuric and sulphurous acids (60 per cent), nitric acid (35 per cent) and hydrochloric acid (5 per cent). Such pollution may affect areas hundreds of kilometres removed from the source of the oxides. Figure 14.11 shows the effects of acid rain at both international and local scales.

Eventually these gases and the particulate matter are removed from the atmosphere by:

1 wet deposition, which itself takes two forms
(a) rainout – the material is incorporated into clouds and lost in raindrops and
(b) washout where material held in suspension below the cloud is swept out by precipitation;
2 dry deposition, which involves gravitational settling and direct impact with terrestrial surfaces.

Figure 14.12 shows the direct and indirect effects of acidification identified by the National Swedish Environmental Protection Board (1983). The fates of the acids inducing these changes are summarized in Figure 14.13. The consequences for the landscape are remarkable. In Sweden, over 18,000 lakes were acidified; about 4000 were severely acidified with severe ecological

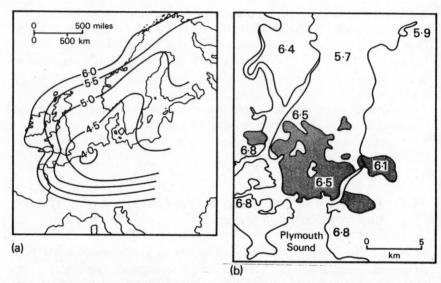

Figure 14.11 (a) acidity (pH) of precipitation in NW Europe in 1966; (b) precipitation acidity (pH) around Plymouth, Devon, England in February 1974 Urban area stippled – wind predominantly from the south-west quadrant. (Redrawn with kind permission of Devon Trust for Nature Conservation.)

implications. The Scandinavians attribute the source of this acidification to the industries and fossil fuel-burning power stations of the United Kingdom.

The immediate biological consequences of acidification are summarized in Figure 14.14. Several effects in lakes merit special emphasis. The lake's content of toxic aluminium becomes enhanced. A variety of other toxic metal ions such as zinc and lead become more readily soluble and consequently are available to be absorbed by the plants and animals in these environments. Consequently species disappear, species diversity soon declines.

In the early stages of acidification many lakes appear to be able to buffer or neutralize the incoming acid products. In soft waters – as long as bicarbonate ions persist – the incoming products are neutralized. When the bicarbonate ions are depleted the 'Threshold of Lake Acidification' is reached, pH declines rapidly and dismal biological consequences follow.

Effects of acidification upon organisms

The tolerance of fish varies between species and sub-species, the physical conditions of the water, and the age and size of the fish. In some cases a sudden change in pH will cause an 'acid shock' resulting in extensive fish mortalities. This effect happens in spring in northern areas with the release of polluted, snow meltwaters. Another process is long term – a more insidious change in pH which may be caused by the build-up from atmospheric inputs. This can affect fish reproduction and may lead to a change in the size and age

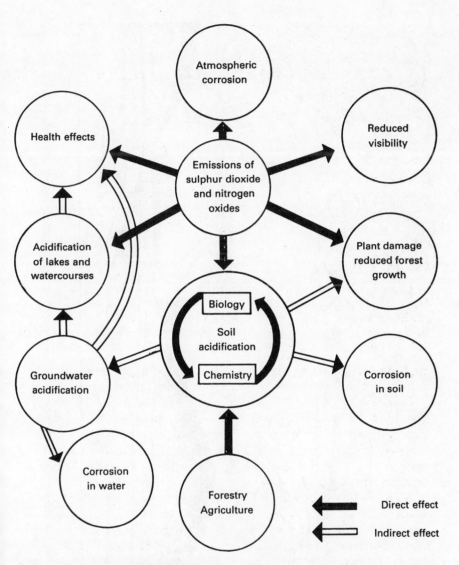

Figure 14.12 Direct and indirect effects of acid rain on ecosystems (After National Swedish Environmental Protection Board (1983).)

of the fish population – proportionally larger and older fish and fewer young, small fish.

Phytoplankton show various changes. Some species are eliminated, whereas others initially undergo major population enhancements – their productivity is also likely to be affected. Some of the semi-aquatic species of *Sphagnum* moss may spread and encroach on to lake margin. Invertebrate populations may become limited in diversity – the number and variety of bottom living species

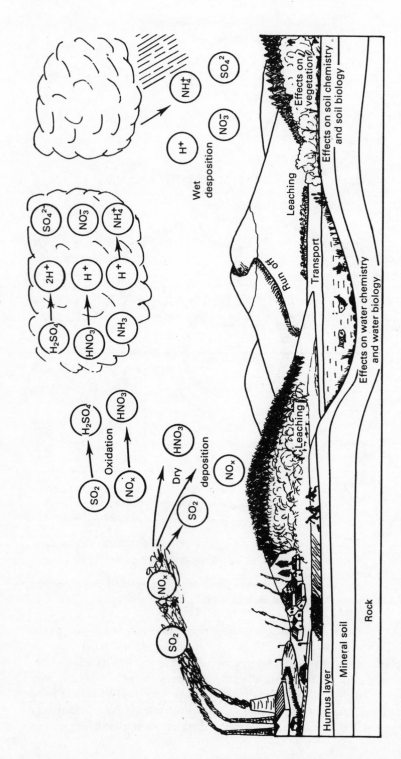

Figure 14.13 Fate of oxides of sulphur and nitrogen in the atmospheric, terrestrial and freshwater environments (After National Swedish Environmental Protection Board (1983).)

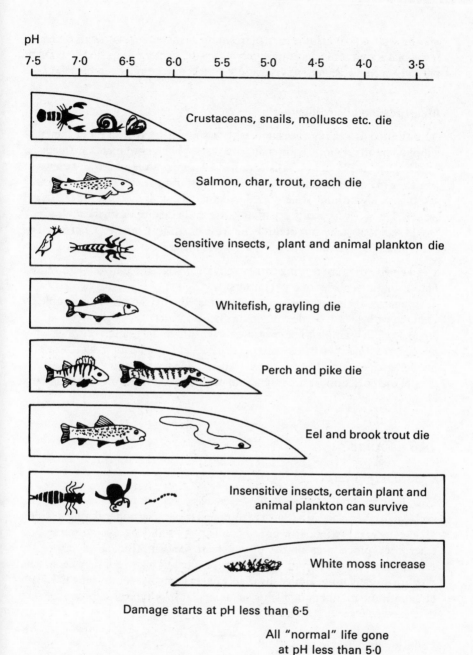

Figure 14.14 Effects of acid rain on species diversity (After National Swedish Environmental Protection Board (1983).)

may be reduced. As acidification of freshwater ecosystems also tends to reduce microbial activity, the decomposition of organic matter is impeded and non or partially decomposed organic detritus can start to accumulate in the lake.

Measurement of acidification

Acidification is readily measured with a simple, inexpensive pH meter as supplied by all scientific equipment suppliers. pH measurements are made on a scale from 1–14. Acid rain is regarded as having a pH of 5.6 or below. In general a pH of 7 is termed neutral, below 7 is termed acid, above 7 is termed alkaline. A logarithmic scale is used to define pH, consequently the fall in pH from 5.7 to 4.7 represents a tenfold increase in acidity, a further fall to 3.7 would represent a one hundredfold increase in acidity from the pH of 5.7. The main precautions to observe when using pH meters are:

1   to ensure they are reading correctly by checking the readings against those of test solutions made from pH buffers;
2   to ensure the electrode is not contaminated with the previous sample by cleaning with a wash bottle containing distilled water;
3   to swirl the sample to be measured to ensure no separation of water types has occurred and then to take two pH readings at 1 minute intervals, these may be averaged, or sampling continued until a stable reading is obtained;
4   pH meter electrodes are fragile and must be carefully protected in and out of use.

14.8   Projects

Classroom projects

If acids or heavy metals are employed in classroom studies it is essential that such work takes place in a properly equipped laboratory under the supervision of a competent scientist who is familiar with the necessary safety and emergency procedures in the event of an accident. Biological laboratory experiments on pollution studies are, as a result of the nature of the phenomena they are investigating, almost invariably likely to involve monitoring the death of significant numbers of plants or animals. Two typical experiments are described below:

Pollution by detergents
1   Prepare 100 ml stock solutions of 1 per cent detergent (by volume or weight). Use this stock to obtain further dilutions of 1:200; 1:500; 1:1000 as required.
2   Add known numbers of *Daphnia* or *Paramecium* from a biological supplier (see Appendix), using a low powered binocular microscope and a small pipette

to obtain a standard volume sample of the 'water', record the numbers of dead individuals every ten minutes for around 3 hours.

3 Plot a curve of the percentage mortality against time and compare the shape of the curves for the various treatments. Describe and discuss your results.

4 Describe and discuss the effects of:

(a) using 'old' *Daphnia*;

(b) carrying out the experiment in a warm and a cold room;

(c) combining the detergent with another pollutant;

(d) changing the pH by adding a weak acid solution;

The effects of acid rain

In the laboratory it is possible to design experiments where the effects of simulated acid rain on invertebrate populations or algae can be investigated. This involves creating samples which differ in their pH and samples which have a standard number of organisms. In addition, it is necessary to be able to monitor the populations of these organisms over a prolonged period.

Particulates

1 Using the methods described in Section 14.5, which indicate how to detect and monitor particulates, compare the nature and extent of damage to the stomata of plants from polluted and non-polluted areas. How far from the likely source are you able to detect particulates in the leaves? Does the diameter of the particles change with increasing distance from the source?

2 Comment on the following:

(a) Is the particulate matter confined to the cuticle?

(b) Are some species more likely, or less likely, to carry particulate matter? Is this propensity related to leaf shape, waxiness, method of leaf presentation, height above ground, extent of pollution?

(c) Does the particulate matter penetrate the stomatal aperture and enter the leaf? Determine the diameter of particles and ascertain the extent to which each size has penetrated?

(d) Does the presence or absence of other vegetation nearby have a scrubbing effect on the air, hence causing less deposition and penetration by particulates?

(e) Does the amount of particulate matter recorded alter after a period of rainfall or a change in wind direction?

(f) Is the vegetation sheltered 'behind' buildings more or less contaminated than vegetation from exposed sites?

Field projects

Wind pruning

The dessicating and abrasive effect of strong winds often causes trees and shrubs to grow in an elongated form downwind. This effect is very noticeable at the coast and in uplands.

1   Using a 1:50,000 map, plot the direction of elongation of trees in an exposed area and identify the extent to which the local wind system is influenced by the local topography.
2   At the coast note the distance inland to which salt-burn effects cause trees and shrubs to lose leaves from their tree crown. Employ the tree crown foliage loss scale described in Section 14.5.
3   Repeat this study around an industrial plant. Look for patterns of leaf loss and, if at the coast, try to identify if the combined effect of possible pollution and salt-enriched onshore winds might be responsible for altering the pattern of crown foliage loss.

Gaseous pollutants
1   Using the procedures given in Section 14.5 determine the concentrations of sulphur dioxide in heavy and light-industrial areas, urban and non-urban areas.
2   Compare sulphur dioxide concentrations near coal burning and gas burning houses.
3   In valleys adjacent to sources of air pollution, identify whether maximum sulphur dioxide concentrations are found on the floor of the valley or higher up on the hillside. Does this change with season, weather conditions, wind strength and direction?
4   Determine whether woodland air contains less sulphur dioxide than non-woodland air in the same region.

Soil pollution and toxicity
1   Obtain a supply of grass seeds from a commercial supplier (see Appendix).
2   Obtain soil samples from spoil tips (you will almost certainly need permission for this), or from locations within and at various distances from polluted ground.
3   Using the bioassay techniques described in Chapter 6, describe and discuss the success of the seeds in each soil type. It will be important to remove the seedlings of competing species whose seeds were in the soil sample.

Freshwater pollution
1   Using information from your local reference library, Environmental Health Officer, water authority or field survey, identify the location of pollution outfalls in streams or rivers in your area. Check and double check with your local authority and the relevant factory management that hazardous effluents are not released.
2   Compare the readings of pH, temperature, dissolved oxygen, flow rate, conductivity, suspended sediments, substrate and marginal vegetation above and below the outfall. Try to identify how these factors alter through the course of the year.
3   Survey the macro-invertebrates, especially the molluscs, above and below the outfall (Chapter 7).

4 Use repeat surveys to identify any longer term changes in the fauna.

A useful method of sampling is to examine a cobble or brick of local material placed on the stream bed. If species cannot be identified, separate the specimens into visually distinct categories. Try to look for trends in shape which will enable you to relate young and mature individuals of the same taxon. Then, count the number of 'types' present and the number of individuals of each 'type' at each location. In particular, note for each taxon: where it lives, how it maintains its position, how it moves, how it feeds, its food, and mode of respiration.

5 Describe and discuss these results.

Acid rain

1 Home-made rain gauges located in back gardens around your local area will enable you to identify the pattern of variation in pH in the local precipitation. Plastic beakers fed by funnels will serve admirably. They must be safe and located above 'dog and cat' height and in the open as far as practicable.

2 Collect the gauges once a week and assess the pH of the precipitation using a pH meter.

3 To what extent do the values vary according to distance from the pollution source, type of pollution source, the time of year, wind direction, precipitation type and quantity, the prevailing weather conditions?

Acid rain is responsible for the accelerated weathering of the stone in many buildings. Identify the age of public buildings by inspection of inscriptions, at the local reference library, or by asking the occupants. Develop a 5 point scale to describe the degree to which the detail of sculpture has been lost and identify the extent to which any patterns of loss reflects distance from a pollution source and/or the age of the building.

The 'black spot' fungal disease of roses is sensitive to atmospheric pollution. Identify the pattern of black spot infestation of garden or park roses. Does this pattern correspond with other evidence of atmospheric pollution? How many gardeners and local authorities are using a fungicide to control the infestation in your area away from pollution sources?

Identify the nature of the correlation between patterns of lichen type, diversity, size and cover, in and around a pollution source yielding sulphur dioxide with observations on acid rain, particulates and tree crown foliage loss. Describe and discuss your findings.

Study design

In all of these projects, it is necessary to pay attention at the design stage to the problems of sampling involved (Section 5.5), the logistics of the exercise, the number of samples that can be analysed in subsequent laboratory study, and the description and statistical methods which will be used to study the field and laboratory data (Chapters 11–13).

15 Biological Conservation

15.1 Conservation in practice

Throughout this book, the examples of differing practical approaches to the study of plants and animals have repeatedly shown the importance of man in both modifying and destroying ecosystems. Often the impacts of people upon wildlife are accidental, unintentional or occur without any consideration or understanding of the biological and ecological systems concerned. However, with advanced intelligence and education, people have become aware of the need for more constructive intervention to modify and control ecosystems, resulting in new branches of ecology and biogeography known as biological conservation and environmental management. A positive attitude to biological conservation accepts that man is inevitably going to adapt and modify ecosystems to suit his own needs and demands. However, modification should minimize the impact on the existing ecosystem and a degree of compatibility must be obtained between people's demands and the long-term status and stability of that ecosystem (Warren and Goldsmith, 1974; 1983).

Sound conservation and management practice requires:
1 an appreciation of the detailed workings of the ecosystem(s) concerned and the likely effects of man's modifications;
2 the collection of data relevant to the specific ecosystem(s) and environmental problem on the basis of which management decisions may be made.

O'Connor (1974) recognized four distinct approaches to conservation practice:

1 species conservation;
2 habitat conservation;
3 conservation as an attitude to land use management;
4 creative conservation.

The following examples, which combine a number of the methods and approaches described in earlier chapters, serve to demonstrate each of these four approaches.

15.2   Species conservation

Much early species conservation was concerned with the protection and preservation of individual rare plant and animal species. Often it was believed that the creation of a nature reserve was sufficient to prevent serious modification or loss of species. However, in a number of examples, such as the Large Copper Butterfly (*Lycaena dispar batavus Obth.*), at Woodwalton Fen National Nature Reserve in Huntingdonshire (Duffey, 1971), this proved inadequate for the survival of the species. Our improved knowledge of ecology has greatly assisted with more recent attempts at the conservation of individual species, but often the picture with respect to a threatened species has been shown to be far more complex than originally realized, particularly in relation to the study of population dynamics, and species life cycle strategies.

The relevance of plant population studies and life cycle strategies to nature conservation

One of the most famous examples of a threat to rare plant species in the British Isles was the proposal in 1965 to construct the Cow Green reservoir in Upper Teesdale within the northern Pennines of England. This was one of the most important botanical sites in Britain. Upper Teesdale contains a number of rare and important arctic-alpine species which are associated closely with the outcrops of sugar-limestone which occur in the valley. When the decision was taken to build the reservoir, a trust fund, Teesdale Trustees (ICI), was set up to sponsor research work on the effects of constructing the reservoir on the vegetation of the floor of the valley. The results of some of this work were published by Bradshaw and Doody (1978).

Flooding of the reservoir occurred during 1970–1, two years after research had started in 1968, and some 8.5 ha of communities containing rare plant species were destroyed. However, populations of several rare species on the adjacent Widdybank and Cronkley Fells were left untouched by the flooding and it was the close proximity of a new large body of water and its effects on these rare species which concerned the conservationists. The researchers realized that if they were to fully understand the ecology of the plant rarities of Teesdale, they had to discover a great deal more about their population dynamics and life cycle strategies.

Sample sites were located on the sugar-limestone soils of both Widdybank and Cronkley Fells. Inert nylon-coated metal marker pegs were sunk into the soil to indicate the corners of permanent quadrats (Section 5.2). Each site could thus be relocated using known distances along a compass bearing from an easily found landmark. Within each quadrat, individual plants of each species had their locations fixed on a superimposed grid and recorded on a chart. Recordings were made at least three times a year of:

1 all individuals on the chart;
2 additions as seedlings or ramets (new units);
3 dead and missing plants;
4 plants with flowers.

Great care was taken not to disturb the site when recording.

Results

One of the primary aims was to assess the size and stability of the plant populations involved. Figure 15.1 shows the populations of selected plant species for the years 1968–76. An important distinction exists between the life cycle strategies and the modes of reproduction of the differing species. The intrinsic rate of natural increase in higher plants has two components, one related to seed production and the other to vegetative reproduction. Examples of species with reproduction only by seed are *Draba incana* (Hoary whitlow grass) and *Polygala amarella* (Limestone milkwort). *Gentiana verna* (Spring gentian) is a perennial species which reproduces by both seed and vegetatively, while *Viola rupestris* x *riviniana* (Teesdale violet) is a sterile perennial which reproduces only vegetatively.

The results of population counts for these species over an eight-year period show clear variations (Figure 15.1). All species showed a dramatic increase in numbers in the autumn of 1969 or spring of 1970 due to particularly favourable growing conditions. However, by 1976, populations of both *Draba incana* and *Polygala amarella* had dropped significantly below the original values of 1968, all showing a steady decline after 1970. In contrast, *Gentiana verna* and *Viola rupestris* × *riviniana* populations remained relatively stable.

As a tentative conclusion, it would appear that those species which were adapted to reproduction by seed, such as *Polygala amarella* and *Draba incana* were in decline and might not be viable into the future. However, those species which reproduced vegetatively appeared to be more stable (*Gentiana verna* and *Viola rupestris* x *riviniana*). This difference may be attributed in part to the management of the Upper Teesdale area for grazing by rabbits, sheep and ponies. Those species with vegetative reproduction at or below ground level are relatively unaffected by grazing, while those reproducing by seed borne on aerial parts of the plant are seriously retarded if grazing is too intensive. Thus, it may be suggested that either protection or relaxation from grazing may be necessary to maintain viable populations of the seed-producing species.

Bradshaw and Doody are, however, cautious about their results. Many practical problems exist with this type of work.

1 The timespan of recording is only eight years and thus it is difficult to know how these initial data relate to any natural population cycle within the plant species. It would appear that prior to 1970, most species populations were at a low point in their cycles and had recording begun in 1972, the dramatic decreases in *Draba* and *Polygala* would have caused considerable alarm.

Population

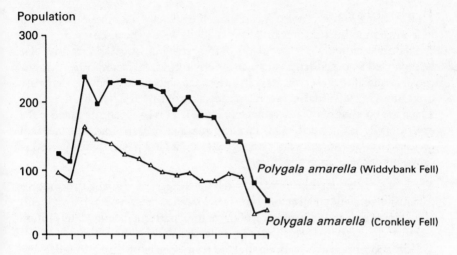

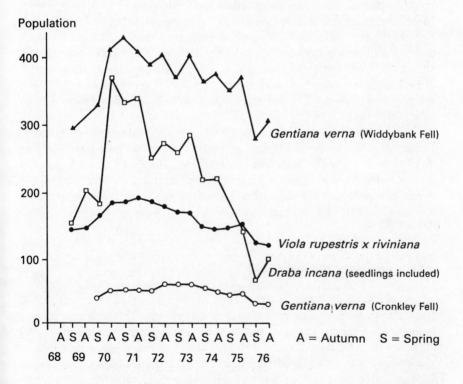

Figure 15.1 Population numbers of four Teesdale species at spring and autumn 1968–76 (Adapted from Bradshaw and Doody (1978); redrawn with kind permission of Applied Science Publishers Ltd.)

2   Recording 2–3 times a year was absolutely essential if data on the longevity of individuals and the turnover of the populations were to be discovered.
3   Natural fluctuations in population numbers of individual species must also be separated from sudden changes due to environmental factors other than the construction of the reservoir. It is an interesting coincidence that the sharp rise in numbers of individuals of several species occurred in 1970, which is the year in which the flooding of the reservoir commenced. This has been attributed to the exceptionally good weather in the early summer of that year, but had that occurred one or two years later, the results might have posed some interesting problems of causal interpretation.

These points demonstrate extremely well the dangers of jumping to erroneous conclusions on only small amounts of data, as well as the considerable problems of obtaining a clear understanding of the nature, performance and likely future viability of rare species which may be under threat.

More recent research work on the rare plant species of Upper Teesdale has involved the use of transplant experiments (Cranston and Valentine, 1983). Ideas concerning transplants have already been introduced in Chapter 1 and in Section 9.5. Cranston and Valentine attempted to preserve by cultivation a number of the rare plants of the Upper Teesdale flora. The plants were removed from their original site and carefully nurtured in the experimental gardens of Manchester University at Jodrell Bank, Cheshire. Some species such as *Viola rupestris x riviniana* and *Draba incana* were very successful. Others, such as *Gentiana verna* were not.

This study raises a number of important philosophical questions about the conservation of rare species. Is there any real purpose to such studies? Cranston and Valentine argue that such transplants are valuable, first to provide material for research work in physiology and genetics and second, culture collections can maintain populations of species whose natural populations have been destroyed and thus make it possible to reintroduce species – a kind of plant zoo.

15.3   Habitat conservation

At a larger scale, biological conservation may be concerned with examples of particular plant communities or complete ecosystems. This is the purpose of habitat conservation. In recent years, increasing concern has been expressed about the need for habitat conservation in highly modified landscapes such as the lowland agricultural landscape of Britain, the Great Plains of the central United States or the large areas of South America formerly occupied by extensive tracts of tropical rain forest.

In 1984, the Nature Conservancy Council of Great Britain published a major report on habitat loss and the need for its conservation. The scale of loss identified is very substantial. As an example, the report states that 95 per cent of lowland herb-rich meadows now lack any significant wildlife interest; 80 per cent of chalk and Jurassic limestone grassland have been lost to arable cultivation or improved grassland since 1940 and 40 per cent of lowland heaths on acidic soils have been destroyed in the same period.

One of the main aims of the conservationists is that sufficient fragments of the former vegetation cover remain in each of these instances.

Island biogeographic theory and habitat conservation

Recent developments in the theory of island biogeography (MacArthur and Wilson, 1967) have led numerous research workers to suggest that the number and size of the remaining fragments as well as the distances between them may be vital in determining the long-term success of habitat conservation. The most fundamental concept of island biogeographic theory is that a relationship exists between the size of a fragment of remaining semi-natural vegetation and the number of species of any group of organisms living within that fragment.

In Britain, relationships for plant species numbers in relation to fragment area have been demonstrated by Usher (1979) and Kent (1972) and Kent and Smart (1981) (Figure 15.2), while relationships for birds (Moore and Hooper, 1975) and butterflies (Shreeve and Mason, 1980) have also been derived. A number of authors have also made recommendations for the conservation of the remaining fragments of the vegetation cover and for the establishment of networks of nature reserves based on these ideas. Of these, the most important are those of Diamond (1975) which are summarized in Figure 15.3.

Clearly the basic principles of habitat conservation in these circumstances are that distances between remaining fragments should be as small as possible, while the size of remaining fragments should be as large as possible. Furthermore, the shape of the remnants should be compact rather than elongated.

A nature conservation review

Another very important aspect of nature conservation must be to maintain the maximum range and diversity of habitat types within a country or region. With this in mind, the Nature Conservancy Council prepared their Nature Conservation Review, which was published in 1977 (Ratcliffe; 1977). The aim of this review was to categorize as fully as possible the outstanding examples of all types of semi-natural vegetation and habitat in the British Isles. The guidelines and principles on which sites were chosen are described in detail in the introduction to the review. Potential sites were described by field workers and then ten factors individually assessed, before being combined subjectively to indicate overall conservation value. These criteria were:

1 extent – importance of a site tends to increase with area;
2 diversity – both species and habitats within the site;

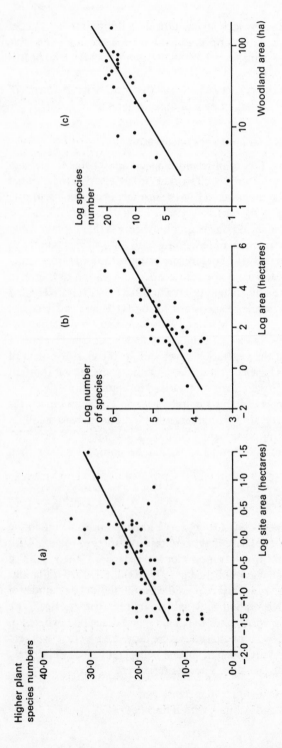

Figure 15.2 Relationships between species numbers and area of fragmented remnant habitats: (a) higher plant species in Dorset (Kent and Smart (1981)); (b) higher plant species on 35 nature reserves in Yorkshire (After Usher (1979); redrawn by kind permission of Blackwell Scientific Publications Ltd.); (c) butterfly species in 22 woods in Shropshire. (After Shreeve and Mason (1980); redrawn by kind permission of Springer-Verlag, Heidelberg.)

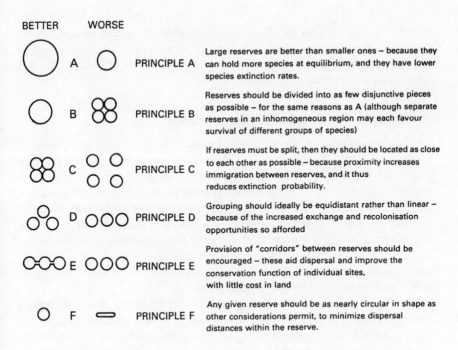

BETTER   WORSE

A   PRINCIPLE A   Large reserves are better than smaller ones – because they can hold more species at equilibrium, and they have lower species extinction rates.

B   PRINCIPLE B   Reserves should be divided into as few disjunctive pieces as possible – for the same reasons as A (although separate reserves in an inhomogeneous region may each favour survival of different groups of species)

C   PRINCIPLE C   If reserves must be split, then they should be located as close to each other as possible – because proximity increases immigration between reserves, and it thus reduces extinction probability.

D   PRINCIPLE D   Grouping should ideally be equidistant rather than linear – because of the increased exchange and recolonisation opportunities so afforded

E   PRINCIPLE E   Provision of "corridors" between reserves should be encouraged – these aid dispersal and improve the conservation function of individual sites, with little cost in land

F   PRINCIPLE F   Any given reserve should be as nearly circular in shape as other considerations permit, to minimize dispersal distances within the reserve.

Figure 15.3 Design principles for nature reserves and for conservation of a fragmented semi-natural vegetation cover in a highly modified landscape (After Diamond (1975); redrawn with kind permission of Applied Science Publishers Ltd.)

3   naturalness – the degree of disturbance or modification to the site by man;
4   rarity – both species and habitat;
5   fragility – sensitivity to disturbance;
6   typicalness – examples of all communities both common and rare must be conserved;
7   position in ecological/geographic unit – proximity to other valuable sites could increase the scientific interest of a site;
8   recorded history – if such information is already available about a site this could be important;
9   potential value – some sites could increase in interest if managed correctly;
10   intrinsic appeal – sites containing particularly distinctive species or of specific interest for some other reason.

Clearly several of these factors overlap and inevitably the final assessment of relative merit will contain a significant element of opinion or subjectivity. Nevertheless the review is a very impressive document and represents a substantial contribution to our knowledge of the range and spectrum of habitat types within the British Isles.

A total of 735 sites were identified as worthy of reserve status in 1977, to which a further 73 sites had been added by 1984. However, a considerable number of these sites have actually been so seriously damaged or destroyed that they are no longer worth conserving. In 1983, there were still only 193 national nature reserves in Britain.

Habitat conservation and ecological evaluation

The fate of wildlife in many areas of the world frequently rests with planning authorities. Often the planners and politicians involved in planning future land use have little or no knowledge of the non-urban environment. The problem is particularly acute in heavily populated and agriculturally advanced areas. A major effect of the ecological movement over the past twenty years has been to increase the general level of awareness in planning circles of the importance of wildlife conservation, not only for its own sake, but also for an influential and often vocal sector of the electorate. Consequently, planners have sought methods to gather 'biological-ecological-environmental-conservation' data as inputs into their planning process.

While planning authorities will generally accept, if not always act upon, arguments concerning sites which contain rare or beautiful species, the identification and designation of larger areas of importance for wildlife which can be satisfactorily managed is usually an extremely difficult task, requiring information of a type which is both difficult and expensive to obtain. Also, convincing the authorities of the need for wildlife conservation at this larger scale is often impossible. However, one answer has been for the ecologist or planner involved to be asked to produce a map which shows their assessment of the relative ecological interest and value of the land in a given planning area. For practical purposes, the ecological categories recognized will have to range from minimal or low value to high, with the field worker knowing full well that only the top one or two categories have any chance of being treated favourably. Also the value of the particular parcel of land will obviously vary according to the size and diversity of the planning region, as well as the size and diversity of the particular habits involved.

The kind of criteria used to decide on the value of a given area of habitat have already been described in the previous section, and study of this list shows the problems involved in attempting to make this kind of assessment. Many of the items must be assessed using different measurement scales and are difficult to combine in any satisfactory manner. Nevertheless, although one may quibble with the quasi-objectivity of indices and numbers representing ecological value which can be generated by assessing such a diverse range of criteria, as long as the subjective nature of the approach remains clear, useful progress can be made.

Ecological evaluation surveys of relatively small areas may provide extremely valuable student projects. However, in their design and execution, several points must be kept firmly in mind.

Table 15.1 *Checklist for a modified form of ecological evaluation devised by Tubbs and Blackwood (1971)*

Ecological zone type occupying 0.5 km² in each 1 km² grid square	Unsown vegetation including non-plantation woodland	Plantation woodland	Agricultural land
Relative value defined by:	Defined as Category I or II depending upon a subjective assessment of rarity of habitat type and the presence of features of outstanding scientific importance	Defined as Cat. II or III, depending upon a subjective assessment of value as wildlife reservoirs	To identify category determine a total score according to the frequency of the following stated desirable features, and compare with guide at bottom left of this table

Scores*

	0	1	2	3
Permanent grassland				✓
Hedgerows and hedgerow timber		✓		
Boundary banks, roadside cuttings, banks, verges				✓
Park timber, orchards, not in commerical prod.	✓			
Ponds, ditches, streams, watercourses				✓
Fragments of unsown vegetation incl. woods smaller than 0.5 km²		✓		
Agricultural land total score	0	2	0	9

Date   4 June 1984
location   Kenn Moor, Avon
Nat Grid ref.   ST 430700

Ecological scoring	Ecological category	Show which category is indicated for this 1 km² square
over 18	I	
15–18	II	
11–14	III	✓
6–10	IV	
0–5	V	

* 0 = none; 1 = present; 2 = numerous; 3 = abundant

1   Are they mainly concerned with a single site evaluation or with relative assessment of a number of sites within an area? In both cases, skills of taxonomy and species identification may be required plus a knowledge of the relative rarity, abundance and representativeness of the species in relation to the county, state or region. The types of maps described in Chapter 9 may be of value here.

2   The exact meaning of 'ecological value' must be discussed and decided among those involved, together with the standards which will enable the workers to identify an ecologically significant area or site. This is usually a controversial part of the project.

3   How can this ecological value, once specified, be measured in a systematic, meaningful and objective manner?

Usually the identification of ecological value depends on the purpose behind a given appraisal. As an example, the South Australian Landsat mapping programme (Section 10.10) is inherently stressing 'naturalness' by mapping the extent of human influence on the structure and composition of the vegetation. Commonly, it is possible to identify two broad categories of standards which have been used in ecological evaluation surveys. The first group include standards for assessing ecological significance and involve a description of the biological components of the area irrespective of the uses to which they are put, or whether they are man-modified or not. The second group of standards have a more man-orientated aim, emphasising the worth, desirability or usefulness of the biological resource to man. In this view, ecological value would include such factors as the value of the biological resource as genetic breeding material, its potential for acting as a habitat for unwanted pest or predator species, research and educational potential and often its aesthetic contribution to the overall landscape.

Once the principal criteria for assessing ecological value are specified, it is then vital to appreciate that the resources of time, money and expertise required for field survey and mapping are often inadequate to complete the task in the most satisfactory manner. Thus the factors noted, measured or mapped must be easily comprehended and recognized by the field worker, readily handled in any subsequent analysis and give results that the sponsoring authority can utilize. As the unusual, unknown or unexpected are almost certain to be amongst the first features encountered during survey, procedures must be as flexible and as universally applicable as possible. The conflicts which clearly emerge in achieving these ideals are self-evident and again will always provide fruitful discussion and argument when such ideas are proposed as a student project.

Subjective assessments of ecological value
A most influential and notably subjective approach to ecological evaluation was published by Tubbs and Blackwood in 1971. This pioneering study was

developed for the structure plan of the county of Hampshire in southern England. The method started by defining three major types of land use and landscape within the county, which were:

1 the complex of habitats comprising agricultural land (zone type 3);
2 plantation woodland (zone type 2);
3 unsown vegetation (zone type 1).

These were considered to represent a sequence of increasing ecological value within the county. Five grades of ecological value were recognized and denoted from I to V with category I of highest value. First, the county was divided up into zones with each zone corresponding to one of the three categories. These were then drawn on a base map. The minimum size for an ecological zone was 0.5 km². Second, the ecological value of each zone was assessed using three main concepts:

1 Unsown or semi-natural habitats have limited distribution in lowland Britain and are subject to pressures from reclamation and development. Hence they have a high conservation value.
2 Areas of plantation woodland often form valuable wildlife reservoirs (thus they also have a relatively high conservation value);
3 Ecological interest in agricultural land will vary inversely with the density and intensity of use of agricultural land.

Thus ecological evaluations for the zones were:

Zone 1: Category I or II (depending on subjective estimates of rarity of habitat type and presence of features of outstanding scientific importance)
Zone 2: Category II or III (based on subjective estimation of the value of the habitat as a wildlife reservoir)
Zone 3: Relative value is a function of habitat diversity and involves the following procedures.

1 Record the presence of habitat features:
(a) permanent grassland;
(b) hedgerows and hedgerow timber;
(c) boundary banks, roadside cuttings, banks and verges;
(d) park timber and orchards;
(e) ponds, ditches, streams and other watercourses;
(f) fragments of other unsown vegetation (including woodland) smaller than 0.5 km².
2 Allocate a score for the presence of each group of features:
0 = none/virtually none in the zone;
1 = present but not a conspicuous feature;
2 = numerous (conspicuous feature);
3 = abundant.

3 Sum the scores for the individual features present in each zone to give a total. Depending on the total, place the zone into one of the evaluation categories as follows:

Total score	Category
15–18	II
11–14	III
6–10	IV
0–5	V

An example of the application of this technique using a standardized checksheet for the extensively drained coastal lowland plain of the Somerset Levels near Bristol, England is shown in Table 15.1. The area under study came within zone type 3 (agricultural land) and thus was scored on the more detailed scale of habitat diversity. The assessment of individual habitats is shown in the right hand column with the summation to a total in the bottom left corner. The total was 11 points indicating that the survey unit should be mapped as category III. If this procedure is repeated within each zone recognized within the region of study, a map of relative ecological evaluation may be generated.

Discussion
Many may be concerned that the method is assessing what cannot be assessed and adding scores that should not really be added. However, if carried out by a person who is well-acquainted with the region, or a group of people who are agreed on their criteria of assessment, then within the area known and studied by that group some useful evaluations may be completed fairly rapidly. However, comparisons with other areas, or with the data of other workers, or modified scoring systems, or for differing environments, will all clearly often be invalid.

15.4   Conservation as an attitude to land use management

Traditionally, biological conservation has been concerned with species and habitat conservation on sites dominated by natural or semi-natural vegetation. However, it is increasingly being realized that a full conservation policy for any developed country must also include land which is primarily in other uses, such as arable farming, grazing management, forestry or recreation. Here there are two aspects to conservation practice. First, there is scope for the maintenance and development of any wildlife interest as a secondary land use. Second, the manner in which the primary land use management practices are carried out can also greatly influence potential wildlife interest in the site itself as well as in adjacent areas which are relatively unmodified. Thus in agriculture, the methods of cultivation, soil management, drainage or fertilizer practice or grazing management in a field are extremely important in maintaining any wildlife interest in that field or its adjacent hedges, woodlands or ponds. Sound land use management is also certain to be of value to the farmer in terms of the long-term yield and return from his agricultural activities which form the primary land use.

Grazing management and species composition on chalk grasslands in Yorkshire

A very simple example of the above principle is demonstrated in Figure 15.4. Shimwell (1973) studied numerous chalk grassland environments in East Yorkshire, many of which were primarily managed as pasture for sheep grazing. He was able to show clearly how the nature of the grazing management influenced the distribution of pasture species in the swards and more particularly how this affected both the value of the swards as food for the grazing animals (the primary land use) and how the species diversity and interest of the swards (the secondary land use) could be maintained.

Six categories of plant life form were recorded as shown in Figure 15.4. The explanation of these categories is as follows:

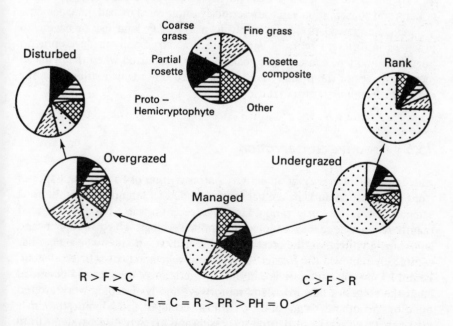

Figure 15.4 Life-form spectra from chalk grasslands in East Yorkshire under different grazing intensities (Shimwell (1973); redrawn with kind permission of the Kent Trust for Nature Conservation.)

1　Coarse grass
2　Fine grasses
3　Rosette-composite plants – species with dense basal rosettes with an aerial part bearing flowers only (e.g. *Leontodon* – Hawkbit)
4　Partial rosette plants – commonly biennials with the largest leaves in a rosette at the base in the first year and then forming on an elongated aerial shoot in the second (e.g. *Ranunculus* – Buttercup)
5　Protohemicryptophytes – species bearing aerial shoots with large aerial leaves and buds protected by small scale-like leaves (e.g. *Hypericum* – St. John's Wort)
6　Others

In the well managed situation (centre – Figure 15.4), both conservation interest, in terms of species diversity and also grazing value, in terms of palatable species, are maximized. However, either over or under-grazing may have very serious consequences for the balance of species. With overgrazing, all categories are substantially reduced, except for the rosette-composite species which are low in palatability. Hence they dominate and simplify the sward. In the opposite case, with under-grazing a different group of plants, the coarse grasses come to dominate and their rapid growth and shading effects quickly eliminate species in all the other five groups. Thus once again, both grazing and conservation value are seriously diminished by bad management.

The same general principles apply to many other land use management practices in different types of agriculture, industry or urban development. Surprisingly often, conservation goals may be maximized when the management practices of the primary land use are also those which will ensure the optimum economic return in the long term.

15.5　Creative conservation

The emphasis of conservation activity on prevention of habitats and species from modification and change has tended to detract attention from the final category of creative conservation. However, recent research has demonstrated that there is ample scope for biologists, applied ecologists/biogeographers and landscape architects in the creation of new habitats in areas where man has completely removed the former semi-natural vegetation cover or ecosystem. Examples of such situations are in the re-vegetation of derelict and degraded land, the verges of new roads and motorways and landscaping work around buildings in both town and country. From an ecologist's standpoint, the aim is usually to try to accelerate natural succession and to re-create ecosystems from scratch as rapidly as possible. Inevitably, the ecosystem which results is usually considerably simplified when compared with the previous semi-natural situation but may nevertheless be of ecological interest and value.

A further very important aspect of many re-vegetation and landscaping projects is that although often very simplified in species composition when the landscaping is first carried out, in time, other species from the surrounding area will invade spontaneously and both flora and fauna will increase in both numbers and diversity. A critical contributory factor will be the management practices which are carried out to maintain that site. The following project demonstrates how management practices must be studied to see how they may influence the species composition of such areas of creative conservation.

The effect of mowing on the distribution of birds in roadside verges

During 1980, the Danish Road Directorate were preparing new instructions for the maintenance of verges along public roads. Previous recommendations dating from 1975 applied only to main highways and required one mowing in autumn. On main roads, mowing is also carried out in summer for traffic safety. For minor roads only general guidelines were available and probably largely ignored.

The aim of the study by Laursen (1981) was to assess the effects of different mowing practices, and particularly summer mowing, on the bird populations of various types of road verge. On the basis of the results, it was intended that recommendations should be forwarded to the Danish Road Directorate for circulation to all local authorities.

The part of the survey which is described here was carried out in 1977 on four stretches of road in East and West Denmark. The distribution of habitats on a typical road verge are shown in Figure 15.5. The whole of the left of each roadside was cut in June/July (summer mowing) while the right was only cut to a width of one passage of the mower, with the remainder uncut to act as a control. Both verges were mown to their full width in the autumn.

Birds were recorded during the period 31 April to 25 July 1977, using

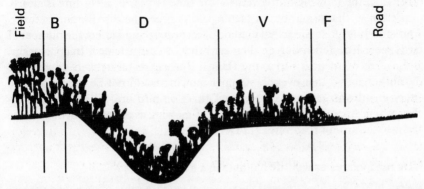

Figure 15.5 The typical structure of a Danish road verge. The roadsides are divided into four sections: **F** front edge; **V** verge; **D** ditch; **B** back edge. (After Laursen (1981); redrawn with kind permission of Applied Science Publishers Ltd.)

a mapping method and transect counts, whereby the number of birds, their location and activities were plotted on detailed maps (See Section 7.9). Each verge was visited four times before and after summer mowing. On each occasion the roadside was walked through at a moderate pace.

Figure 15.6 shows the variations in distribution of the main groups of bird species before and after summer mowing. Before mowing, the skylark (*Alauda arvensis*) nested in the back edge (B) and foraged on the verge (V) or front edge (F), while the corn bunting (*Emberiza calandra*) preferred the back edge (B) and the ditch (D). The sparrows (*Passer domesticus* and *Passer montanus*) clearly liked the back edge for nesting (B) and the front edge for foraging (F).

After mowing the patterns change comparatively little for the skylark, although there is an increased preference for the back edge (B). The bunting, however, showed a slightly increased preference for the back edge (B) and avoided the front (mown) edge completely. Sparrows and linnets (*Carduelis cannabina*) showed the greatest change both with marked increase of activity in the front edge (F) where they could find insects killed by cars. The linnets were apparently not present before mowing. Thus both sparrows and linnets dramatically changed their foraging habits before and after mowing.

However, the most important result of the survey related to the total numbers of each species recorded before and after mowing. From Figure 15.6, it is apparent that skylark numbers fell only slightly, corn bunting numbers by nearly half and other species by a third, while house and tree sparrow numbers increased threefold and linnets arrived on the mown verge. Chi-squared tests (Section 12.4) showed that overall there was no significant difference in the distributions before and after mowing. Thus, although changes had occurred in the types of species and their foraging habits, overall there was no substantial loss of interest in bird life.

Once again, there are clearly problems in this type of work. The changes observed, even though not significantly different, may be related to factors other than the mowing of the verge – the time of year of recording is clearly important in that a number of factors could affect the distribution between April and July. Also the actual method of recording could affect the number of birds recorded, depending on their sensitivity to disturbance. In conclusion, as far as recommendations to the Danish Road Directorate were concerned, the author found that overall, summer mowing could not be shown to have either an advantageous or detrimental effect on bird life. This conclusion is clearly very important, given the extent of mowing as a management practice and also the area of road verges in Denmark.

New road verges as wildlife habitats in a new city

Kelcey (1975) has also stressed the potential wildlife value of new road verges. These may be summarized as follows.

1   Unmanaged grassland is becoming increasingly scarce with ever-increas-

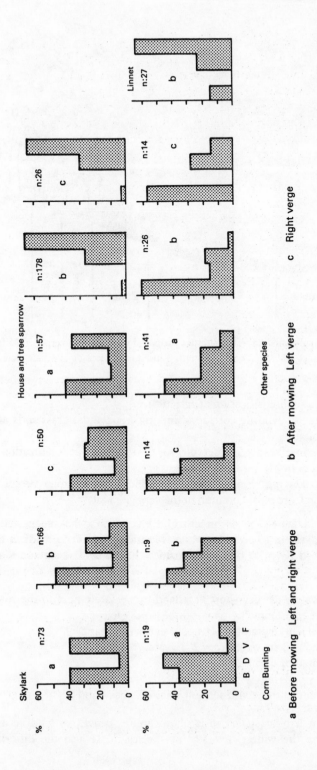

Figure 15.6 Variations in total numbers and the micro-habitat distribution of birds on Danish road verges, before and after summer mowing (see text) (After Laursen (1981); redrawn with kind permission of Applied Science Publishers Ltd.)

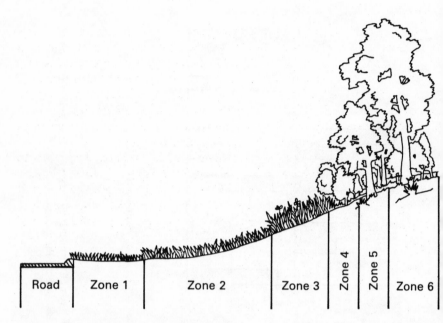

Figure 15.7   The six-zoned habitat structure proposed for new road verges in Milton Keynes (After Kelcey (1975); redrawn with kind permission of Elsevier Science Publishers BV.)

ing urbanisation and intensification of agriculture.

2   Road-sides act as 'wildlife corridors' linking woodlands, grasslands and other similar habitats.

3   Verges can themselves act as reservoirs for the spontaneous colonization of nearby disturbed or newly landscaped areas.

4   According to Perring (1967), over 700 species grow on road verges in Britain, 50 of which are most common in the verge habitat.

As an exercise in landscape architecture, Kelcey (1975) has suggested a modified zoning system to provide habitat variation adjacent to roads in the new city of Milton Keynes, Buckinghamshire, England. The six zones are (Figure 15.7):

Zone 1 – 1–2 m strip of very short grassland immediately next to the road, maintained by frequent mowing two or three times each year.

Zone 2 – medium length grassland cut once every 2–3 years.

Zone 3 – tall grassland cut once every 2–3 years.

Zone 4 – dominated by shrubs.

Zone 5 – largely small trees with some shrubs.

Zone 6 – largely forest trees with some species of zones 4 and 5.

This variation in habitat structure is of maximum benefit to wildlife and should encourage the widest diversity of species. Note also the differing frequency and intensity of management and maintenance practices. Other problems which are raised are the actual choice of species to be planted in each zone and what should be the balance of native and non-native species? Kelcey concludes that there are many conflicts involved and one of the most crucial problems is the lack of fundamental quantitative ecological data about the wildlife value of individual species and tree/shrub associations.

The problems and projects described in this last chapter have demonstrated the way in which some of the techniques of applied ecology and biogeography may be put to practical use in providing data and information of relevance to applied studies and environmental management. Numerous potential projects and exercises exist both in this applied area and in the many more academic topics covered earlier. It is to be hoped that the information in this text may help towards an increased understanding of the principles and practice of biogeography and ecology and in particular, show how relevant projects may be planned and executed effectively.

15.6 Projects

Classroom projects

Ecological evaluation of alternative new road routes from Clarendon to the Mount Bold Reservoir, South Australia

Data sources
(a) 1:50,000 topographic series: sheet Noarlunga, South Australia
(b) 1:10,000 orthophotomap sheet No.6627–12
(c) 1:10,000 black and white aerial photograph cover Clarendon to Mount Bold Reservoir (optional)

Background information
An (imaginary) highway authority has decided that a new first-class dual carriageway road is to be constructed from Clarendon to the main dam of the Mount Bold Reservoir (Figure 15.8). This route would be used primarily by heavy construction vehicles travelling to the dam and while on subsequent maintenance. Once built, the dam would be visited by tourists in cars wishing to see the reservoir and to 'recreate' there. To minimize other environmental disruption, all new road improvement and widening will take place only where a track or route already exists. (A short cut of up to 200 metres across fields/paddocks is permitted).

There are obvious requirements for the route:

1   the route should be as short as possible;
2   gradients should be as gentle as possible;
3   there should be as little earthmoving as possible;
4   there should be as little ecological damage as possible.

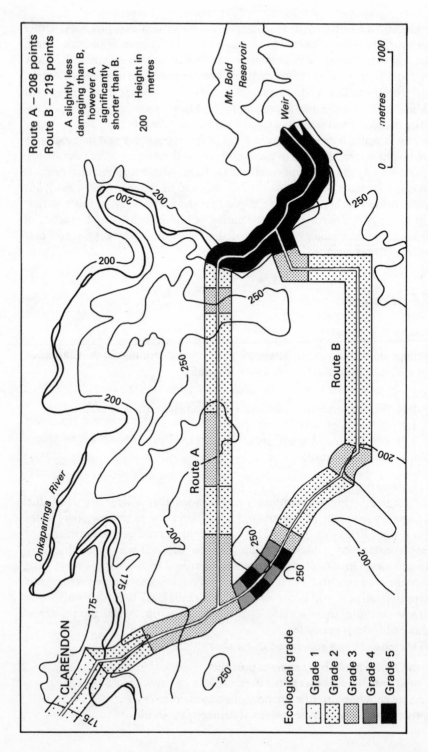

Route A – 208 points
Route B – 219 points

A slightly less damaging than B, however A significantly shorter than B.

200   Height in metres

Mt. Bold Reservoir

Weir

Onkaparinga River

CLARENDON

Route A

Route B

Ecological grade
Grade 1
Grade 2
Grade 3
Grade 4
Grade 5

0   1000
:metres

Figure 15.8   The ecological evaluation of two first class (imaginary) roads from Clarendon to Mount Bold Reservoir, South Australia

Examine the topographic map and decide which routes best meet criteria 1, 2 and 3.

Ecological evaluation
The approach used here is that of Tubbs and Blackwood (1971) but modified for problems such as this by Yapp (1973) and further adapted for use in this exercise.

The following rating scale for ecological evaluation was devised for the Clarendon area.

Grade V land   Unsown natural vegetation. If woodland, then tree cover over 50 per cent; if wetland, then undrained.
Grade IV land   Semi-natural habitats – unplanted woodland where cover less than 50 per cent or partly drained wetlands.
Grade III land   Plantations, orchards or agricultural land of some diversity.
Grade II land   Agricultural land of low diversity.
Grade I land   Ecologically valueless land – mostly urban.

An estimate of the ecological value of each proposed route may be rapidly obtained using the above valuation:

1   Take a 100 m strip on either side of the proposed routes.
2   Divide the strip into areas of grades I–V using information from maps and photos.
3   Multiply the total length of grade V land by 5; grade IV by 4, grade III by 3; grade II by 2 and grade I by 1.
4   Add the total ecological values obtained from each proposed (imaginary) route.

The route with the lowest sum represents that with the least ecological value and hence the least environmental impact. The resulting evaluation map is shown in Figure 15.8 for two routes.

Although this project has been described for an area in South Australia, the ideas are very easily adaptable for any location where good base maps and preferably aerial photography are available. Indeed, the exercise is made all the more valuable because of the adaptations necessary to apply the ideas in different situations.

Financial complications
Further variations on this exercise include specifying very low maximum road gradients of 1:30, or indicating a nominal cost of construction per km (e.g. A$200,000 per km) for land purchase, compensation and actual construction costs. These data permit comparisons of the financial implications of differing routes, which may then be compared with the ecological evaluations. Interesting situations arise when the cheapest route emerges as also being the ecologically most valuable and sensitive. How can these problems be resolved?

Field projects

Permanent quadrats and plant population numbers on road verges

1   Read Sections 5.2 and 15.2.

2   This is a long-term project to be carried out over a number of years.

3   Locate several differing areas of semi-natural grassland on road verges – try to find verges of differing ages.

4   Set up several permanent quadrats (1 m^2) on each area by positioning small marker posts at each corner.

5   Count the numbers of individuals of each species and map their position in each quadrat. Note that this cannot be done for many species of grass or moss since it is impossible to define individuals in many cases. Cover estimates may be substituted.

6   Also record the vigour of each species.

7   Return and repeat the measurements 2–3 times per year at different times in the growing season over a number of years.

8   Plot the results as a graph of numbers of individuals of each plant species over time, or changes in percentage cover values. Comment on changes in relative numbers of plants growing and on new colonizers and extinctions. Are the verges becoming more diversified in their flora or more simplified? If any of the verges were landscaped or seeded with a commercial seed mix, what is the balance of those seed mix species with new arrivals from the local flora?

9   If new species do invade over the study period, look in the adjacent area for the habitats from which they may have dispersed.

Species–area relationships and island biogeographic theory

1   Read the introductory material on island biogeography and habitats (Section 15.3).

2   In Section 15.3, linear relationships were shown between species numbers and area for fragments of remnant semi-natural vegetation within highly modified agricultural landscapes. Note the nature of the relationships in Figure 15.2. For graph (a), the area is expressed in logarithms while the higher plant species numbers are on an arithmetic (normal) scale. However, graphs (b) and (c) each have both axes in logarithms. Ensure you are familiar with the meaning of logarithmetic scales.

3   Locate an area of intensive agriculture in your home region.

4   Test the above relationships to see if they apply in your vicinity.

Equipment

A base map at an appropriate scale – preferably with woodland and other semi-natural vegetation shown.

Graph paper (fine grid) or a planimeter (for measurement of areas).

Procedure

1 On the base map, delinate a sample region containing 20–40 areas of remnant vegetation cover.

2 Assuming that the work is to be completed in an agricultural area, check with landowners about access and request permission if necessary.

3 In the field, check or mark on the base map, the location of each of the remnant areas. Try to do this as accurately as possible since it is these areas which will be used to determine the species–area relationships.

4 Visit each site of semi-natural vegetation at the timne of year when the flora/fauna are best developed. Visit more than once if necessary.

5 Try to walk around the whole area of each site, counting the numbers of species of the group(s) of plants or animals in which you may be interested. Species may be identified, but this is not essential.

6 On return to the laboratory, measure the area of the remnant fragments off the base map using either a planimeter or by counting the number of grid squares covered when a remnant area on the map is superimposed over graph paper. Then convert the grid square values to the number of grid squares occupying a known area on the map, to derive actual area figures.

7 Draw graphs of species numbers in relation to site area for:

(a) arithmetic species v arithmetic area;

(b) logarithmic species v logarithmic area;

(c) arithmetic species v logarithmic area.

8 Calculate a Spearman's rank correlation coefficient (Section 13.2) for each of the graphs between species numbers and area. If the relationship is significant, fit a semi-averages regression line to the data (Section 13.4).

9 Comment on your results and in particular, think about variables other than simply size and area of site which could determine variations in species numbers in your region.

10 Generate new hypotheses concerning these other factors and devise further projects to test them.

Glossary

Arresting factor   An edaphic, topographic or biotic factor which holds a plant community in an apparently stable state during succession. When the factor is removed succession will usually move on to the climatic climax.

Autecology   The study of an individual organism or species in relation to its environment including its life-history and behaviour.

Autotroph   An organism that assimilates energy from either sunlight (green plants) or inorganic compounds. See also Heterotroph.

Biogeocoenosis   An ecosystem comprising a definite, limited and homogenous community e.g. forest stand, lake, etc. Term very common in Europe, N. America, Australasia but not in the UK.

Biogeochemical cycle   Circulation of chemical elements between the biotic and abiotic components of the biosphere.

Biogeography   The description and explanation of distributions of the biological resources of the Earth's surface involving a full understanding of organism environment relationships.

Biomass   The weight of living material in all or part of an organism, population or ecosystem. Usually expressed as dry matter/unit area, e.g. kg/ha or g/m^2.

Biome   A major regional ecological community of plants and animals extending over a large area, with a uniform climate and vegetation.

Biota   Groups of plants and animals living together in one place.

Biotic effects   Any influences on an ecosystem which result from the activities of living organisms including man.

Climatic climax   The final stage of a succession where the dominant plant community is in equilibrium with the prevailing climate.

Climax   The final stage of the successional sequence or sere which consists of a relatively stable ecosystem which is in equilibrium with its environment.

Commensalism   A feeding relationship whereby one partner benefits but the other is unharmed.

Community   A group of organisms inhabiting a common environment and usually interrelated by food chains.

Competition   The process by which species with similar requirements fight for resources such as water, light, food or space.

Descriptive statistics   Methods of statistical analysis used to describe the nature and variability of individual variables.

Detritus food chain   The food chain which exists among organisms in the soil which relies on organic matter and litter from the grazing food chain for its energy input.

Disclimax   An apparently stable climax where a rapidly spreading new species has invaded and become dominant. In reality it is inherently unstable.

Diversity   The number and variety of species in a community (species richness).

Ecological evaluation   The assessment of biological resources in terms of their interest and value, usually for planning purposes.

Ecological niche   The status or position occupied by an organism in a community in terms of its structural, physiological or behavioural adaptions to its immediate environment and other organisms.

Ecology   The study of the relationships of plants and animals to each other and to their environment.

Ecosystem   Any identifiable unit in nature consisting of both living organisms and their physico-chemical environment in which there is a continuous exchange of inorganic materials between the living organisms and their environment.

Ecotone   The boundary or transitional zone between two types of ecosystems.

Edaphic factors   Factors influencing ecosystem and community status which are attributable to soil conditions.

Energy flows   The pathways of energy derived from solar radiation through trophic levels in an ecosystem. At each transfer some energy is degraded as heat.

Environmental gradient   The continuum of one environmental factor ranging between extremes, e.g. a gradation from hot to cold; exposed to sheltered; wet to dry.

Eurytypic species   A species which has a very wide range of environmental tolerance and hence will tend to be widely distributed.

Eutrophication   Nutrient enrichment of bodies of water caused by organic enrichment, e.g. sewage disposal or runoff from heavily fertilized agricultural land.

Evolution   The process of gradual and continual change in the life-form or behaviour of the successive generations of an organism.

Factor compensation   The substitution of one environmental factor which may be in abundance for another which may be deficient, thus allowing a species to survive beyond the limits of the deficient factor.

Fauna   The animal species of a given region or time period.

Flora   The plant species of a given region or time period.

Flora(s)   A hierarchical classification of the plants of a region based on physiological and evolutionary characteristics, also used by botanists for identification purposes.

Floristic data   Data collected on plant species composition and also often abundance.

Food chain   A simple linear feeding relationship between organisms at successive trophic levels.

Food web   A complex of feeding relationships between species at successive levels.

Gaussian curve (normal curve)   The bell-shaped curve which commonly results from the plotting of the frequency of occurrence of a species across the complete range of one environmental gradient.

Grazing food chain   The food chain between plants, herbivores and carnivores which exists above ground. On death the tissues of these organisms provide the energy input for the detritus food chain.

Gross primary production   The total energy assimilated by plants and other photosynthetic organisms in a given area over a period of time.

Habitat   The place where a plant or animal normally lives, characterized by its physical characteristics, e.g. a lake or a forest.

Habitat classification   The recognition of habitats of different types and characteristics.

Halosere   Primary vegetation succession in a saline environment e.g. a salt marsh.

Herbivore   An organism which consumes living plants or their parts.

Heterotroph   An organism that utilizes organic materials as a source of energy and nutrients.

Homeostatic mechanism   The means by which an ecosystem maintains internal stability in the face of a fluctuating external environment.

Hydrosere   Primary vegetation succession in a freshwater or waterlogged environment e.g. a lake edge or river margin.

Hypothesis   A provisional explanation of observed facts which is usually then tested for its validity.

Inferential statistics   A set of formal statistical procedures which enable an investigator to test hypotheses through analysis of relevant data and which also allows him/her to be relatively precise about his degree of uncertainty concerning the result.

Latin binomial   The internationally recognized system of species nomenclature. Every plant or organism has two Latin names (hence bionominal) where the first refers to the genus to which the species belongs and the second refers to the species within that genus.

Leaching   Removal of soil materials in solution by percolating waters.

Limiting factors   An environmental factor limiting the growth or reproduction of an individual or community.

Macro-nutrient   A nutrient which is required by plants in relatively large quantities e.g. nitrogen, phosphorus and potassium.

Master limiting factor   The one limiting factor required by a species which is not present in sufficient amounts to allow the species to survive.

Micro-climate   The climate of a small localized area which is usually modified by the presence of vegetation, buildings, land use or topographic features.

Micro-habitat   The particular parts of the habitat that an individual organism encounters in the course of its activities.

Micro-nutrient   A nutrient which is required by plants in small amounts, but which is nevertheless essential to its growth and survival.

Net primary production   The total energy assimilated by plants and other photosynthetic organisms minus that lost in respiration, in a given area over a period of time.

Null hypothesis   The statement of a hypothesis as a negative proposition, formulated for the purpose of applying an inferential statistical test to the problem under investigation.

Nutrient   Any substance required by organisms for normal growth and maintenance. (Mineral nutrients usually refer to inorganic substances taken up by plants from soil or water.)

Omnivore   An organism which eats both plants and animals.

pH   A measure of the acidity of a solution pH2–5 = acid, pH5–7 = neutral up to pH8–14 = alkaline; defined as the logarithm to the base 10 of the reciprocal of the hydrogen ion concentration.

Photosynthesis Utilization of the energy of light to combine carbon dioxide and water into simple sugars.

Physiognomy The physical structure or life form of an organism.

Plagioclimax A stage of vegetative succession which is kept at apparent equilibrium or stability by the activities of man or animals in burning, clearing or grazing.

Plant community An assemblage of plant species which tend to grow together under similar environmental and/or management conditions.

Plasmolysis The removal of water from a plant cell by osmosis with resultant shrinking.

Population The total number of organisms of a particular taxonomic group inhabiting a particular area.

Population dynamics Changes in the numbers of individuals of a particular taxon through time.

Prisere The sequence of changes in vegetation and associated ecosystem characteristics on a new, fresh, biologically unmodified site, ending in a stable state or climax.

Profile diagram A diagram, usually drawn to scale, depicting the vertical structure of vegetation. Dominant species are usually identified.

Quadrat A sampling unit of varying size for studying and measuring vegetation. Traditionally quadrats are represented as square areas.

Respiration Metabolic assimilation of oxygen accompanied by release of energy, carbon dioxide and water, and by breaking down of organic compounds.

Secondary succession The sequence of vegetation and ecosystem changes on a site which was formerly vegetated or which has suffered vegetation modification to an earlier seral stage.

Sere See Succession

Species A group of actually or potentially interbreeding populations that are reproductively isolated from all other kinds of organisms.

Stability The inherent capacity of any system to resist change.

Stenotypic Species which has a very narrow range of environmental tolerance and hence will tend to be narrowly distributed.

Succession The regular and progressive change in the components of an ecosystem from the initial colonization of an area to a stable state or mature ecosystem.

Synecology The study of groups of organisms associated as a unit and their relationships to one another and to their environment.

Taxon Taxonomic unit of any scale.

Tolerance The ability of an organism to endure a prevailing environmental condition.

Trophic level Position in the food chain determined by the number of energy-transfer steps to that level.

Vegetation Plants collectively or plants in the mass.

Vegetation formation type A primary spatial subdivision of the vegetation of the Earth based largely on life-form and physiognomy of the vegetation e.g. tropical rain forest, savannas, temperate forest, tundra. (See Biome).

References

Alcock, C. R. and Symon, D. E. (1977), *The Flora*, in Gilbertson, D. D. and Foale, M. R., *The Southern Coorong and Lower Younghusband Peninsula of South Australia*, Adelaide, Nature Conservation Society of South Australia, pp. 25–38

Alder, H. L. and Roessler, E. B. (1964), *Introduction to Probability and Statistics*, San Francisco: Freeman

Alderdice, D. F. (1967), 'The detection and measurement of water pollution – biological assays', *Canadian Department of Fisheries; Canadian Fish Report*, **9**, pp. 33–9

Alvin, L. and Kershaw, K. A. (1963), *The Observer's Book of Lichens*, London: Warne

American Public Health Association, American Waterworks Association and Water Pollution Control Federation (1976), *Standard methods for the examination of water and wastewater*, 14th edn, Washington DC: APHA

Auble, D. (1953), 'Extended tables for the Mann-Whitney statistic', *Bulletin of the Institute of Educational Research at Indiana University*, **1**, no. 2

Bang, P. and Dahlstrom, B. (1974), *Collins Guide to Animal Signs*, translated by G. Vevers, London: Collins

Bannister, P. (1976), *Physiological Plant Ecology*, Oxford: Blackwell Scientific Publications

Barkham, J. P. (1978), 'Pedunculate oak woodland in a severe environment: Black Tor Copse, Dartmoor', *Journal of Ecology*, **66**, pp. 707–40

Barrington, E. T. W. (1980), *Environmental Biology*, London: Arnold

Bayfield, N. G. (1971), 'A simple method for detecting variations in walker pressure laterally across paths', *Journal of Applied Ecology*, **8**, pp. 533–5

Beals, E. W. and Cope, J. B. (1964), 'Vegetation and soils in eastern Indiana woods', *Ecology*, **45**, pp. 777–92

Benson, L. (1969), *The Cacti of Arizona* (3rd ed), Tucson: University of Arizona Press

Berry, R. J. (1970), 'The natural history of the house mouse', *Field Studies*, **3**, pp. 219–62

Boycott, A. E. (1934), 'The habitats of land Mollusca in Britain', *Journal of Ecology*, **22**, pp. 1–38

Boycott, A. E. (1936), 'The habitats of freshwater Mollusca in Britain', *Journal of Animal Ecology*, **5**, pp. 116–86

Boyden, C. R. (1972), 'Relation of size to age in the cockles *Cerastoderma edule* and *Cerastoderma glaucum* in the River Crouch estuary, Essex', *Journal of Conchology, London* **27**, pp. 475–89

Bradshaw, M. E. and Doody, J. P. (1978), 'Plant population studies and their relevance to nature conservation', *Biological Conservation*, **14**, pp. 223–42

Brandt, C. J. and Rhoades, R. W. (1973), 'Effects of limestone dust on lateral growth of forest trees', *Environmental Pollution* 4(3), pp. 207–8

Bridges, E. M. (1978), *World Soils*, (2nd edn), London: Cambridge University Press

Briggs, D. J. (1977), *Sources and Methods in Geography: Sediments*, London: Butterworths

Briggs, D. J. (1977), *Sources and Methods in Geography: Soils*, London: Butterworths

Briggs, D. J. (1983), *Biomass Potential of the European Community*, Sheffield: University of Sheffield on behalf of the Environment and Consumer Protection Service, European Economic Community

Briggs, D. J. and Courtney, F. C. (1985), *Agriculture and Environment. The Temperate Agriculture Systems*, London: Longmans

Briggs, D. J. and Smithson, P. A. (1985), *Fundamentals of Physical Geography*, London: Hutchinson

Brockie, R. (1960), 'Road mortality of the hedgehog (*Erinaceous europaeus* L.) in New Zealand', *Proceedings of the Zoological Society of London*, 134, pp. 505–8

Bruun, B. (1975), *The Hamlyn Guide to the Birds of Britain and Europe*, London: Hamlyn

Cameron, R. A. D. and Redfern, M. (1976), 'British Land Snails. Mollusca: Gastropoda', *Synopses of the British Fauna*, no. 6, *Linnean Society of London*, London: Academic Press

Carnahan, J. A. (1976), 'Natural Vegetation', *The Atlas of Australian Resources* (2nd edn) Canberra: ACI, Department of National Resources

Caughley, G. (1977), *Analysis of Vertebrate Populations*, Chichester: Wiley

Chamberlin, T. C. (1965), 'The method of multiple working hypotheses', *Journal of Geology*, 5, pp. 837–48

Chapman, H. D. and Pratt, P. F. (1961), *Methods for the Analysis of Soils, Plants and Waters*, California: University of California, Division of Agricultural Sciences

Christian, C. S. and Perry, R. A. (1953), 'The systematic description of plant communities by the use of symbols', *Journal of Ecology*, 41, pp. 100–5

Clapham, A. R., Tutin, T. G. and Warburg, E. F. (1962), *Flora of the British Isles*, (2nd edn), London: Cambridge University Press

Clapham, A. R., Tutin, T. G. and Warburg, E. F. (1981), *Excursion Flora of the British Isles*, (3rd edn), Cambridge: Cambridge University Press

Clapham, W. B. (Jr) (1983), *Natural Ecosystems*, New York: Macmillan

Corbett, G. B. and Southern, H. N. (eds.) (1971), *A Handbook of British Mammals*, Oxford: Blackwell Scientific Publications

Cousens, J. (1974), *An Introduction to Woodland Ecology*, Edinburgh: Oliver and Boyd

Cox, B. C. and Moore, P. D. (1985), *Biogeography: an Ecological and Evolutionary Approach*, (4th edn), Oxford: Blackwell Scientific Publications

Cranston, D. M. and Valentine, D. M. (1983), 'Transplant experiments on rare species from upper Teesdale', *Biological Conservation*, 26, pp. 175–91

Crothers, J. H. (1964), 'On the distribution of some common animals and plants along the rocky shores of west Somerset', *Field Studies*, 4, pp. 369–89

Daines, R. H., Motto, H. and Chilko, D. M. (1970), 'Atmospheric lead; its relationship to traffic volume and proximity to highways', *Environmental Science and Technology*, 4, pp. 318–22

Dalby, D. H. (1970), 'The salt marshes of Milford Haven, Pembrokeshire', *Field Studies* 3, pp. 297–330

Dandy, J. E. (1969), *Watsonian Vice-Counties of Great Britain*, London: Ray Society

Dansereau, P. (1957), *Biogeography: an Ecological Perspective*, New York: The Ronald Press

Davies, J. L. (1957), 'A hedgehog road mortality index', *Proceedings of the Zoological Society of London*, **128**, pp. 606–8

Dempster, J. P. (1971), 'Some effects of grazing on the population ecology of the cinnabar moth', in Duffey, E. and Watt, A. S. (eds), *The Scientific Management of Animal and Plant Communities for Conservation*, Oxford: Blackwell Scientific Publications, pp. 517–26

Diamond, J. M. (1975), 'The island dilemma: lessons of modern biogeographic studies for the design of natural reserves', *Biological Conservation*, **7**, pp. 129–46

Dorsett, J. E., Gilbertson, D. D., Hunt, C. O. and Barker, G. W. W. (1984), 'The UNESCO Libyan Valleys Survey IX. Image Analysis of Landsat Data and its Application to Environmental and Archaeological Surveys', *Libyan Studies* **15**, pp. 71–80

Dowdeswell, W. H. (1984), *Ecology; Principles and Practice*, London: Heinemann

Duffey, E. (1971), 'The management of Woodwalton Fen; a multidisciplinary approach', in Duffey, E. and Watt, A. S. (eds.), *The Scientific Management of Animal and Plant Communities for Conservation*, Oxford: Blackwell Scientific Publications, pp. 581–98

Ellis, A. E. (1951), 'Census of the distribution of the British non-marine Mollusca: 7th edition', *Journal of Conchology, London*, **23**, pp. 171–244

Elton, C. S. (1958), *Ecology of Invasions by Plants and Animals*, London: Methuen

Elton, C. S. (1966), *The Pattern of Animal Communities*, London: Methuen

Elton, C. S. and Miller, R. S. (1954), 'The ecological survey of animal communities with a practical system of classifying habitats by structural characters', *Journal of Ecology*, **42**, pp. 460–96

Evans, J. G. (1972), *Land Snails in Archaeology*, London: Seminar Press

Ewen, A. H. and Prime, C. T. (1975), *Ray's Flora of Cambridgeshire*, Hitchin, Herts: Wheldon and Wesley

Fenner, F. (1971), 'Evolution in action: myxomatosis in the Australian wild rabbit', in Kramer, A (ed.), *Topics in the Study of Life*, New York: Harper and Row

Ferry, B. W., Baddeley, M. S. and Hawksworth, D. L. (eds.) (1973), *Air Pollution and Lichens*, London: The Athlone Press, University of London

Findlay, W. P. K. (1975), *Timber: properties and uses*, London: Granada Publishing

Fisher, R. A. and Yates, F. A. (1974), *Statistical Tables for Biological, Agricultural and Medical Research*, (6th edn), Longman: London

Gilbert, O. L. (1965), 'Lichens as indicators of air pollution in the Tyne valley', in Goodman, G. T., Edwards, R. W. and Lambert, J. M. (eds.), *Ecology and the Industrial Society*, Oxford: Blackwell Scientific Publications, pp. 35–48

Gilbertson, D. D. (1975/6), 'The pattern of precipitation acidification around Plymouth, Devon. 1. Trial survey of precipitation in February 1974', *Quarterly Journal of the Devon Trust for Nature Conservation*, **7**, no. 1, pp. 22–4

Gilbertson, D. D. (1983), 'The impact of off-road vehicles in the Coorong lake and dune complex of South Australia', in Webb, R. H. and Wilshire, H. G., (eds.), *The Environmental Effects of Off-Road Vehicles: Impacts and Management in Arid Lands*, Berlin and New York: Springer-Verlag

Gilbertson, D. D. and Foale, M. R. (1977), *The Southern Coorong and Lower Younghusband Peninsula of South Australia*, Adelaide: Nature Conservation Society of South Australia

Gilbertson, D. D. and Pyatt, F. B. (1980), 'Tree crown foliage loss: a mapping study on the Cumbrian coast', *Journal of Biological Education*, **14**, no. 4, pp. 311–17

Gimingham, C. H. (1953), 'Contributions to the maritime ecology of St. Cyrus, Kincardineshire. Part III. The Salt Marsh', *Transaction of the Botanical Society of Edinburgh*, **XXXVI**, no. 2, pp. 137–64

Glater, R. A. B. and Hernandez, L. (1972), 'Lead detection in living plant tissue using a new histochemical method', *Air Pollution Control Association Journal*, 22, no. 6, pp. 463–67

Glue, D. (1969), *Collecting and analysing bird pellets* Reprinted from *Bird Study*. Available from the British Trust for Ornithology, Beech Grove, Tring, Hertfordshire HP23 5NR, UK.

Godwin, H. (1975), *The History of the British Flora: a Factual Basis for Phytogeography*, London: Cambridge University Press

Hammond, R. and McCullagh, P. S. (1978), *Quantitative Techniques in Geography: An Introduction* (2nd edn), Oxford: Clarendon Press

Harris, J. R. and Haney, T. G. (1973), 'Techniques of oblique aerial photography of agricultural field trials', Canberra, *Division of Soils Technical Paper*, no. 19, Commonwealth Scientifica and Industrial Research Organisation

Harrison, R. M. and Laxen, D. P. H. (1978), 'Sink processes for tetra alkyl lead compounds in the atmosphere', *Environmental Science and Technology* 12, pp. 1384–92

Haslam, S. (1978), *River Plants*, London: Cambridge University Press

Haugen, A. D. (1944), 'Highway mortality of wildlife in southern Michigan', *Journal of Mammology*, 25, pp. 177–84

Hawksworth, D. L. (1974), *The Changing Flora and Fauna of the British Isles*, London: Academic Press

Hawksworth, D. L. and Rose, F. (1976), 'Lichens as Pollution Monitors', *Studies in Biology*, no. 66, London: Arnold

Herter, K. (1965), *The Hedgehog*, London: Phoenix Press

Heyligers, P. C., Laut, P. and Margules, P. (1978), 'Vegetative cover and land use for the agricultural districts of South Australia', Canberra, *Technical Memoir* 78/1, CSIRO Division of Land Use Research

Hooper, M. D. (1970), 'The botanical importance of our hedgerows. The Flora of a Changing Britain', *Botanical Society of the British Isles*, **Report 11**, pp. 58–62

Hooper, M. D. (1970), 'Dating hedges', Area 4, pp. 63–5

Howard, J. A. (1970), *Aerial Photo-Ecology*, London: Faber and Faber

Hubbard, C. E. (1968), *Grasses*, Harmondsworth: Pelican

Hubbard, C. E. (1984), *Grasses*, (3rd edn), Harmondsworth: Pelican

Hunter, R. F. (1958), 'Hill land improvement', *Advancement of Science*, London, 15, no. 38, pp. 194–96

Institute of Terrestrial Ecology (ITE) (1977), *Overlays of environmental and other factors for use with Biological Records Centre distribution maps*, 68 Hills Road, Cambridge, CB2 1LA., UK, ITE/NERC

Iversen, J. (1944), '*Viscum, Hedera* and *Ilex* as climatic indicators', *Geologiska Foreningen I Stockholm Fordhandlinger*, 66, pp. 463–83

Janus, H. (1965), *The Young Specialist Looks at Land and Freshwater Molluscs*, London: Burke

Jermyn, S. T. (1974), *Flora of Essex*, Colchester: Essex Naturalists Trust

Kelcey, J. G. (1975), 'Opportunities for wildlife habitats on road verges in a new city', *Urban Ecology*, 1, pp. 271–84

Kent, M. (1972), 'A method for the survey and classification of marginal land in agricultural landscapes', London, *University College, Discussion Papers in Conservation*, no. 1

Kent, M. (1980), 'Regional assessment of plant growth problems for colliery spoil reclamation: I. Introduction and site survey', *Minerals and the Environment*, 2, pp. 165–75

Kent, M. (1982), 'Plant growth problems in colliery spoil reclamation – a review', *Applied Geography*, 2, pp. 83–107

Kent, M. and Smart, N. (1981), 'A method for habitat assessment in agricultural landscapes', *Applied Geography*, 1, pp. 9–30

Kerney, M. P. (1976), *Atlas of the Non-Marine Mollusca of the British Isles*, Cambridge: NERC and the Conchological Society of Great Britain and Ireland

Kerney, M. P. (1976), 'European distribution maps of *Pomatias elegans* (Müller), *Discus ruderatus* (Ferussac), *Eobania vermiculata* (Müller), and *Margarita margaritifera* (Linné)', *Archiv. Molluskenk*, 106, pp. 243–49

Kerney, M. P. and Cameron, R. A. D. (1979), *A Field Guide to the Land Snails of Britain and North-West Europe*, London: Collins

Kershaw, K. A. (1973), *Quantitative and Dynamic Plant Ecology (2nd edn)*, London: Arnold

Knight, J. R. (1978), 'Victoria's forests and man – historical outline', *Victorian Yearbook*, pp. 1–38

Kodak (1977), *An introduction to infra-red photography*, Hemel Hempstead, UK: Kodak Information Sheet AM 900 (H)

Krebs, C. J. (1978), *Ecology: the Experimental Analysis of Distribution and Abundance*, (2nd edn), New York: Harper and Row

Küchler, A. W. (1965), *International bibliography of vegetation maps. Volume 1, North America*, Kansas: University of Kansas Library Series

Küchler, A. W. (1966), *International bibliography of vegetation maps. Volume 2, Europe*, Kansas: University of Kansas Library Series

Küchler, A. W. (1969), *International bibliography of vegetation maps. Volume 3, USSR, Asia and Australia*, Kansas: University of Kansas Library Series

Küchler, A. W. (1970), *International bibliography of vegetation maps. Volume 4, Africa, South America, and the World (General)*, Kansas: University of Kansas Library Series

Lancashire Naturalist's Trust (1975/1976), *Report on the Wyre Estuary – Shard Bridge at Knott End*, Preston: Dr G. Steed, School of Education, Chorley Campus, Preston Polytechnic, Union Street, Chorley RP7 1ED

Laursen, K. (1981), 'Birds on roadside verges and the effects of mowing on frequency and distribution', *Biological Conservation*, 20, pp. 59–68

Laut, P., Heyligers, P. C., Keig, G., Loffler, E., Margules, C., Scott, R. M., Sullivan, M. E. and Lazarides, M. (1977), *Environments of South Australia*, Handbook and 7 volumes, Canberra: CSIRO Division of Land Use Research

Laut, P. (1978), *Supplementary maps for 'Environments of South Australia'*, Canberra: CSIRO Division of Land Use Research, Technical Memorandum 78/8

Laut, P. and Paine, T. A. (1982), 'A step towards an objective procedure for land classification and mapping', *Applied Geography*, 2, 109–26

Lawrence, M. J. and Brown, R. W. (1967), *Mammals of Britain: their Tracks, Trails and Signs*, London: Blandford Press

Lever, C. (1977), *The Naturalised Animals of the British Isles*, London: Hutchinson

Lewis, J. R. (1964), *The Biology of Rocky Shores*, London: English Universities Press

Lines, R. and Howell, R. S. (1963), 'The use of flags to estimate the relative exposure of trial plantations', *Forestry Commission Forest Record*, 51, pp. 1–31

Little, P. and Martin, M. H. (1972), 'A survey of zinc, lead, and cadmium in soil and natural vegetation around a smelting complex', *Environmental Pollution*, 3, pp. 241–54

Locke, G. M. (1962), 'A sample survey of field and other boundaries in Great Britain', *Quarterly Journal of Forestry*, 56, pp. 137–44

Lowe, C. R. and Campbell, H. (1968), 'Bronchitis and atmospheric pollution in two steel works', *Joint meeting on Occupational hazards in Bronchitis*, Cardiff

Lyons, T. R. and Avery, T. E. (1977), *Remote Sensing. A Handbook for Archaeologists and Cultural Resource Managers*, Washington, DC: Cultural Resources Management Division, National Park Service, US Department of the Interior

MacArthur, R. H. and Wilson, E. O. (1967), *The Theory of Island Biogeography*, Princeton, NJ: Princeton University Press

Maitland, P. S. (1972), *Key to British Freshwater Fishes*, Ambleside, Windermere, England: Freshwater Biological Association

Massey, C. I. (1972), 'A study of hedgehog road mortality in the Scarborough district 1966–1971', *The Naturalist*, 97 pp. 103–5

McLeese, D. W. (1956), 'Effects of temperature, salinity and oxygen on the survival of the American lobster', *Journal of the Fish Research Board of Canada*, 13, no. 2, pp. 247–72

Messenger, G. (1971), *Flora of Rutland*, Leicester: Leicester Museums

Middleton, J. T., Emik, L. O. and Taylor, O. K. (1965), 'Air quality and standards for agriculture', *Air Pollution Control Association Journal*, 15, pp. 476–80

Mitchell, B., Staines, B. W. and Welch, D. (1977), *Ecology of Red Deer*, Cambridge: Institute of Terrestrial Ecology, NERC

Mitchell, J. (1973), 'The bracken problem', in Tivy, J. (ed.), *The Organic Resources of Scotland: their Nature and Evaluation*, Edinburgh: Oliver and Boyd, pp. 99–108

Mobberley, D. G. (1956), 'Taxonomy and distribution of the genus *Spartina*', *Iowa State College Journal of Science*, 30, pp. 471–574

Moore, N. W. and Hooper, M. D. (1975), 'On the number of bird species in British woods', *Biological Conservation*, 8, pp. 239–50

Moran, J. M., Morgan, M. D. and Wiersma, J. H. (1980), *Introduction to Environmental Science*, New York: Freeman

Morgan, R. A. (1975) 'The selection and sampling of timber from archaeological sites for identification and tree ring analysis', *Journal of Archaeological Science*, 2, pp. 221–30

Muus, B. J. and Dahlstrom, P. (1972), *Collins Guide to Freshwater Fishes of Britain and Europe*, London: Collins

National Swedish Environmental Protection Board, (1983), *Acidification: a Boundless Threat to Our Environment*, Solna: Sweden

Nature Conservancy Council (1984), *Nature Conservation in Great Britain*, Shrewsbury: Nature Conservancy Council

Norton, R. L. (1982), 'Assessment of pollution loads to the North Sea', *Water Research Centre Report*, no. 233, Medmenham: UK

O'Connor, F. B. (1974), 'The ecological basis for conservation', in Warren, A. and Goldsmith, F. B., (eds.), *Conservation in Practice*, Chichester: Wiley, pp. 87–98

Odum, E. P. (1971), *Fundamentals of Ecology*, (3rd edn), Philadelphia: W. B. Saunders

Odum, E. P. (1983), *Basic Ecology*, Philadelphia: Holt Saunders

Oertli, J. J. (1956), 'Effects of salinity on susceptibility of sunflower plants to smog', *Soil Science*, **87**, pp. 249–51

Öklund, J. (1964), 'The eutrophic Lake Borrevan (Norway) – an ecological study of shore and bottom fauna with special reference to gastropods, including a hydrographic survey', *Folia Limnologika Scandinavia*, **13**, pp. 1–337

Olds, E. G. (1938), 'Distributions of sums of squares of rank differences for small numbers of individuals', *Annals of Mathematical Statistics*, **9**, pp. 133–48

Olds, E. G. (1949), 'The 5% significance levels for sums of squares of rank differences and a correction', *Annals of Mathematical Statistics*, **20**, pp. 117–18

Owen, D. F. (1980), *What is Ecology?*, (2nd edn), London: Oxford University Press

Paijmans, K. (1975), *Explanatory notes to the vegetation map of Papua New Guinea*, Canberra: CSIRO Land Research Series, 35

Patterson, C. C. and Salvia, J. P. (1968), 'Lead in the modern environment', *Environment*, **10**, p. 72

Pears, N. V. (1977), *Basic Biogeography*, London: Longmans

Pearsall, W. H. (1918), 'The aquatic and marsh vegetation of Esthwaite water', *Journal of Ecology*, **8**, pp. 180–202

Perring, F. H. (1967), 'Verges are vital – a botanist looks at our roadsides', *Journal of the Institute of Highway Engineers*, **14**, pp. 13–16

Perring, F. H. and Walters, S. M. (1962), *Atlas of the British Flora*, London: Thomas Nelson and Sons

Peterken, G. F. (1967), 'Guide to checksheet for IBP areas', *IBP Handbook*, no. 4, Oxford: Blackwell Scientific Publications

Phillips, R. (1977), *Wild flowers of Britain*, London: Ward Lock

Phillips, R. (1978), *Trees in Britain, Europe and North America*, London: Ward Lock and Pan Books

Phillips, R. (1980), *Grasses, Ferns, Mosses and Lichens of Great Britain and Ireland*, London: Ward Lock and Pan Books

Pollard, E., Hooper, M. and Moore, N. W. (1974), *Hedges*, London: Collins

Praeger, R. L. (1906), 'A simple method of representing geographical distribution', *Irish Naturalist*, **17**, pp. 88–94

Putnam, R. D. and Wratten, S. D. (1984), *Principles of Ecology*, London: Croom Helm

Pyatt, F. B. (1970), 'Lichens as indicators of air pollution in a steel producing town', *Environmental Pollution*, **1**, pp. 45–56

Pyatt, F. B. (1973), 'Some aspects of plant contamination by airborne particulate matter', *International Journal of Environmental Studies*, **5**, pp. 215–20

Pyatt, F. B. (1974), 'Plant sulphur content as an indication of air pollution from nineteenth century smelting operations', Proceedings of the Second International Conference on Bioindicators of Landscape Deterioration: Ustav Krajinne Ekologie Ceskoslovenske Akademie Ved. Most, Czechoslovakia

Rackham, O. (1975), *Hayley Wood: its History and Ecology*, Cambridge: Cambridgeshire and Isle of Ely Naturalists Trust

Ranwell, D. S. (1967), 'World resources of *Spartina townsendii* (sensu lato) and the economic use of *Spartina* marshland', *Journal of Applied Ecology*, **4**, pp. 239–56

Ranwell, D. S. (1972), *Ecology of Salt Marshes and Sand Dunes*, London: Chapman Hall

Ratcliffe, D. N. (1977), *A Nature Conservation Review*, 2 vols, London: Cambridge University Press

Ray, John, (1660), *Catalogus Plantarum circa Cantabrigium nascentium*, Cambridge: John Field. Reprinted in translation as Ewen, A. H. and Prime, C. T. (1975) (eds.), *John Ray's flora of Cambridgeshire*, Hitchen, Herts: Wheldon and Wesley

Richardson, J. A. (1957), 'Derelict pit heaps and their vegetation', *Planning Outlook*, **4**, pp. 15–22

Ricketts, H. (1966), *Wild Flowers of North America*, 6 vols., New York: McGraw Hill

Ricklefs, R. E. (1979), *Ecology*, (2nd edn), London: Nelson

Riney, T. (1960), 'A technique for assessing physical condition of some ungulates', *Journal of Wildlife Management*, **24**, pp. 92–4

Robbins, R. G., Haantgens, H. A., Mabbutt, J. A., Pullen, R., Reiner, E., Saunders, J. A. and Short, K. (1976), *Lands of the Ramu-Madung area, Papua New Guinea*, Canberra, CSIRO Land Research Series 37

Roebuck, W. D. and Boycott, A. E. (1921), 'Census of the distribution of British land and freshwater Mollusca', *Journal of Conchology*, London, **16**, pp. 165–212

Rose, F. (1981), *Wild Flower Key: British Isles – North West Europe*, London: Warne

Rossin, A. C., Sterritt, R. M. and Lester, J. N. (1983), 'The influence of flow conditions on the removal of heavy metals in the primary sedimentation process', *Water, Air and Soil Pollution*, **19**, pp. 105–21

Royal Commission on Environmental Pollution (1983), *Lead in the Environment*, Ninth report, London: HMSO

Sayer, J. A. (1969), *Some aspects of the management of grazing animals on Dartmoor*, Unpublished MSc. Dissertation, University College, London

Sheail, J. (1980), *Historical Ecology*, Cambridge: ITE

Shimwell, D. W. (1973), 'An introduction to the geography and ecology of chalk grassland', in Kent Trust for Nature Conservation, *Chalk Grassland – studies on its conservation and management*, Maidstone, Kent: Kent Trust for Nature Conservation, pp. 1–5

Shreeve, T. G. and Mason, C. F. (1980), 'The number of butterfly species in woodlands', *Oecologia*, **45**, pp. 414–18

Siegel, S. (1956), *Non-parametric Statistics for the Social Sciences*, New York and Tokyo: McGraw-Hill/Kogakusha

Sládeček, V. (1979), 'Continental systems for the assessment of river water quality', in James, A. and Evison, L., (eds.), *Biological Indicators of water quality*, Chichester: Wiley, 3.1–3.32

Slatter, R. J. (1978), 'Ecological effects of trampling on sand dune vegetation'. Journal of Biological Education, **12**, pp. 81–96

Smith, R. J. and Atkinson, K. (1975), *Techniques in Pedology: a Handbook for Environmental and Resource Studies*, London: Elek Science

Smith, I. and Lyle, A. (1979), *Distribution of Freshwaters in Great Britain*, Cambridge: Institute of Terrestrial Ecology, NERC, 68 Hills Road, Cambridge CB2 1LA

Soil Survey of England and Wales (1975), *Soil Map of England and Wales, 1:1,000,000*, Harpenden: UK

South Australian Department of Lands (1973), *Vegetation and range site plan of the arid zone*, Adelaide: compiled by the Pastoral Board as the basis of an inventory of CSIRO research into pastoral ecosystems of the arid zone. Government Printer, Adelaide

South Australian State Planning Authority (1970), *South East Planning Area development Plan*, Adelaide: Government Printer

South Australian State Planning Authority (1970), *Kangaroo Island Planning Area Development Plan*, Adelaide: Government Printer

South Australian State Planning Authority (1974), *Mount Lofty Ranges Study*, Adelaide: Government Printer

South Australian State Planning Authority (1975), *Outer Metropolitan Planning Area Development Plan*, Adelaide: Government Printer

South Australian State Planning Authority (1977), *Riverland Planning Area Development Plan*, Adelaide: Government Printer

Specht, R. L. (1972), *The Vegetation of South Australia*, (2nd edn), Adelaide: Government Printer

Specht, R. L., Roe, E. M. and Boughton, V. H. (1974), 'Conservation of major plant communities in Australia and Papua New Guinea', *Australian Journal of Botany*, **Supplement 7**

Spence, D. H. N. (1964), 'The macrophytic vegetation of freshwater lochs, swamps and associated fens', in Burnett, J. H. (ed.), *The Vegetation of Scotland*, Edinburgh: Oliver and Boyd, pp. 306–425

Sprague, J. B. (1969), 'Measurement of pollutant toxicity to fish, I. Bioassay methods for acute toxicity', *Water Research*, **3**, pp. 793–821

Steenbergh, W. F. and Lowe, C. H. (1977), *Ecology of the Saguaro: II Reproduction, Germination, Establishment, Growth and the Survival of the Young Plant*, Washington, DC: National Parks Service Monograph Series no. 8

Takala, K. and Olkkonen, (1981), 'Lead content of an epiphytic lichen in the urban area of Kuopio, east central Finland', *Annales Botanici Fennici*, **18**, no. 2, pp. 85–9

Tansley, A. G. (1935), 'The use and abuse of vegetational concepts and terms', *Ecology*, **16**, pp. 284–307

Tansley, A. G. (1939), *The British Islands and their Vegetation*, London: Cambridge University Press; 2 volumes, reprinted in 1965

Teather, E. K. (1970), 'The hedgerow: an analysis of a changing landscape feature', *Geography*, **55**, pp. 146–55

Thomas, M. D. (1958), 'Air pollution with relation to agronomic crops. I. General status of research on the effects of air pollution on plants', *Agronomy Journal*, **50**, pp. 545–50

Thomas, M. D. (1961), 'Effects of air pollution on plants', in *World Health Organisation, Air Pollution*, Geneva: WHO

Thomas, M. D. and Hill, G. R. (1935), 'Absorption of sulphur dioxide by alfalfa and its relation to leaf injury', *Plant Physiology*, **10**, pp. 291–307

Thomas, M. D. and Hendricks, R. H. (1956), 'Effect of air pollution on plants', in Magill, P. L., Holder, H. R. and Ackley, C. *Air Pollution Handbook*, New York: McGraw-Hill

Thornton, I. (1980), 'Geochemical aspects of heavy metal pollution and agriculture in England and Wales', in Ministry of Agriculture Fish and Food, *Inorganic Pollution and Agriculture Reference Book*, London: HMSO

Tivy, J. (ed) (1973), *The Organic Resources of Scotland*, Edinburgh: Oliver and Boyd

Tivy, J. (1982), *Biogeography: a Study of Plants in the Ecosphere*, (2nd edn), Edinburgh: Oliver and Boyd

Tomlinson, R. W. (1973), *The Inyanga area: an essay in regional biogeography*, Harare: University of Zimbabwe (formerly Rhodesia) Occasional Paper no. 1

Tubbs, C. R. and Blackwood, J. W. (1971), 'Ecological evaluation of land for planning purposes', *Biological Conservation*, **3**, pp. 169–72

Twidale, C. R., Tyler, M. J. and Webb, B. P. (1976), *Natural History of the Adelaide Region*, Adelaide: Royal Society of South Australia

Usher, M. B. (1979), 'Changes in the species-area relations of higher plants on nature reserves', *Journal of Applied Ecology*, **16**, pp. 213–15

Victorian Department of Mines (1866), *Victoria: distribution of forest trees*, Map, Publications Section, Department of Minerals and Energy, 107 Russell Street, Melbourne 3000, Australia

Walter, H. (1973), *Vegetation of the Earth: in Relation to Climate and Eco-physiological Conditions*, London: English Universities Press/Springer-Verlag

Ward, S. D., Jones, A. D. and Manton, M. (1972), 'The vegetation of Dartmoor', *Field Studies*, 3, pp. 505–33

Warner, C. G., Davies, G. M., Jones, J. G. and Lowe, C. R. (1969), 'Bronchitis in two integrated steel works, II Sulphur dioxide and particulate pollution in and around the works', *Annals of Occupational Hygiene*, 12, pp. 151–70

Warren, A. and Goldsmith F. B. (eds.) (1974), *Conservation in Practice*, Chichester: Wiley

Warren, A. and Goldsmith, F. B. (eds.) (1983), *Conservation in Perspective*, Chichester: Wiley

Watson, H. C. (1847–59), *Cybele Britannica*, London: Longmans. (supplement in 1860, compendium 1870)

Watson, H. C. (1873–74), *Topographical Botany*, London: Ditton, 2nd edn, 1883, Quaritch

White, R. E. (1979), *Introduction to the Principles and Practice of Soil Science*, Oxford: Blackwell Scientific Publications

Whiteley, D. and Yalden, D. W. (1976), 'Small mammals in the Peak District', *The Naturalist*, 101, pp. 89–101

Williams, M. (1974), *The making of the South Australian landscape: a study in the historical geography of Australia*, London: Academic Press

Woodell, S. R. J., Mooney, H. and Hill, A. J. (1969), 'The behaviour of *Larrea divaricata* (creosote bush) in response to rainfall in California', *Journal of Ecology*, 57, pp. 37–45

Woodiwiss, F. S. (1964), 'The biological system of stream classification used by the Trent River Board', *Chemistry and Industry*, 11, pp. 443–47

Yalden, D. W. (1977), *The identification of remains in owl pellets*, Reading: UK, Occasional Publications of the Mammal Society, Harvest House, 62 London Road, Reading

Yapp, W. B. (1973), 'Ecological evaluation of a linear landscape', *Biological Conservation*, 5, pp. 45–7

Young, H. E. and Kramer, P. J. (1952), 'The effect of pruning on the height and diameter growth of loblolly pine', *Journal of Forestry*, 50, no. 6, pp. 474–79

Zimmerman, P. W. and Hitchcock, A. E. (1956), 'Susceptibility of plants to hydrofluoric acid and sulphur dioxide gases', *Contributions of the Boyce Thompson Institute*, 18, pp. 263–79

Appendix

Field safety

Natural Environment Research Council, Polaris House, North Star Avenue, Swindon SN2 1ET, *Safety in the Field*.
British Mountaineering Council, Crawford House, Precinct Centre, Booth St., East Manchester M13 9RZ *Safety on Mountains*.

Equipment suppliers and other useful addresses

Advisory Centre for Education, 32 Trumpington St., Cambridge, England, for an excellent clean air guide and lichen identification chart.
British Ecological Society, Harvest House, 62 London Road, Reading RB1 5AS.
BDH Chemicals Ltd, Freshwater Road, Dagenham, Essex, RM8 1RZ.
British Trust for Ornithology, Beech Grove, Tring, Herts. HP23 5NR.
Griffin and George Ltd, Ealing Road, Alperton, Wembley Middx. HA0 1HJ, for scientific and survey equipment for schools and colleges.
Mammal Society, Harvest House, 62 London Road, Reading. RG 5AS.
Philip Harris Biological Ltd, Oldmixon, Weston-Super-Mare, Avon, for biological and pollution study materials.
Science Studios (Oxford), 7 Little Clarendon St., Oxford OX1 2HP for Beck hand microscopes.
Stanton Hope Ltd, 11 Seax Court, Southfield, Basildon, Essex, for Pressler Tree Corers.
J. Stanier and Co., Manchester Wireworks, Sherborne St., Manchester 3, for nylon mesh of fine grades.

Sources of maps, aerial photographs and satellite imagery

Map suppliers
Geocenter, Postfach 8008 30, D-7000, Stuttgart 80, West Germany.
Geocenter, Lutzourstrasse 105/106, Berlin 30, West Germany.
Hammond Map Store Inc., 12 East 41st St., New York, NY 10017, USA.
Institut Geographique Nationale, 136 bis rue de Grenelle, 75700 Paris, France.
Stanford's International Map Centre, 12–14 Longacre, London WC2 2LP, UK.

Australia
Maps and atlas of Australian resources: Division of National Mapping, P.O. Box 548, Queanbyan, NSW 2620.
Air photographs and Landsat imagery: Air Photographs Pty Ltd, 620–624 Burwood Road, Auburn, Vic. 3123.
South Australia: Surveyor General's Office, Department of Lands, PO Box 1266, Adelaide, South Australia 5001.
Western Australia: Vegetation Survey of Western Australia – series of seven maps and accompanying memoirs, University of Western Australia Press, Nedlands, Perth, WA 6009.
Vegmap Publications, 1:250,000 sheets, 6 Fraser Road, Applecross. WA 6153.

Canada
National Air Photo Library, Surveys and Mapping Building, 615 Booth St., Ottawa, Ontario K1A 0E9.
Map Distribution Office, Survey and Mapping Branch, Department of Mines and Technical Services, 615 Booth St., Ottawa, Ontario K1A 0E9.

France
Institut Geographique National, 136 bis rue de Grenelle, 75700 Paris.
Spot satellite imagery – GIFAS, Service Press-Information, 4 rue Galilee, 75116, Paris.
Spot data facilitates mapping at 1;100,000 scales and can also be viewed stereoscopically.

Papua New Guinea
Surveyor General, National Mapping Bureau, Department of Lands, Surveys and Environment, PO Box 5665, Boroko, PNG.

United Kingdom
Ordnance Survey for topographic and specialist maps. Romsey Road, Maybush, Southampton SO9 4DH.
The Librarian, Soil Survey of England and Wales, Rothamstead Experimental Station, Harpenden, Hertfordshire AL5 2JA – soil maps and memoirs.
The Biological Records Centre, Monks Wood Experimental Station, Abbots Ripton, Huntingdon and the Institute of Terrestrial Ecology, 68 Hills Road, Cambridge CB2 1LA for biological surveys and environmental overlays.
Land Use maps, Miss Alice Coleman, Department of Geography, Kings College, University of London, The Strand, London WC2 2LS.
Directorate of Overseas Surveys, Kingston Road, Surbiton, Surrey KT5 9TS for studies and maps of smaller countries, especially commonwealth countries.

Aerial photograph suppliers
Department of Environment, Air Photograph Unit, Prince Consort House, Albert Embankment, London SE1 7TF, for a comprehensive index of available aerial photographs in the United Kingdom.
Aerofilms Ltd, Gate Studios, Station Road, Borehamwood, Berks. WD6 1EG.
Meridian Airmaps Ltd, Marlborough Road, Lancing, Sussex BN15 8TH.
Hunting Surveys Ltd, Elstree Way, Borehamwood, Herts WD6 15B.
Clyde Surveys (formerly Fairey Surveys), Reform Road, Maidenhead, Berks. SL6 8BU.

United States of America
Aerial Photographs – apply for the *Status of Aerial Photography* index at the Map Information Office, US Department of the Interior, Geological Survey, National Center, Reston, Virginia 22092, USA.
Satellite and high altitude imagery (Landsat and ERTS), EROS Data Center, Sioux Falls, South Dakota 57198.
Agricultural stabilization and conservation service, USDA, Aerial Photography Field Office, 2505 Parleys Way, Salt Lake City, Utah 84109.
Forest Service, US Department of Agriculture, Washington, DC 20250.

Animal identification guides

Bang, P. and Dahlstrom, B. (1974), *Collins Guide to Animal Tracks and Signs*, London: Collins

Bruun, B. (1975), *The Hamlyn Guide to the Birds of Britain and Europe*, London: Hamlyn
Cameron, R. A. D. and Redfern, M. (1976), 'British land Snails. Mollusca: Gastropoda', *Synopses of the British Fauna*, no. 6, *Linnean Society of London*, London: Academic Press
Corbett, C. B., and Southern, H. N. (eds.) (1971), *A Handbook of British Mammals*, Oxford: Blackwell Scientific Publications
Janus, H. (1965), *The Young Specialist Looks at Land and Freshwater Molluscs*, London: Burke
Kerney, M. P., and Cameron, R. A. D. (1979), *A Field Guide to the Land Snails of Britain and North West Europe*, London: Collins
Lawrence, M. J. and Brown, R. W. (1967), *Mammals of Britain: their tracks, trails and signs*, London: Blandford Press
Lever, C. (1977), *The Naturalised Animals of the British Isles*, London: Hutchinson
Maitland, P. S. (1972), *Key to British Freshwater Fishes*, Ambleside: Windermere, Freshwater Biological Association
Muus, B. J., and Dahlstrom, P. S. (1972), *Collins Guide to the Freshwater Fishes of Europe*, London: Collins

Plant identification guides

Alvin, L. and Kershaw, K. A. (1963), *The Observer's Book of Lichens*, London: Warne.
Fitter, R. S. R. and McClintock (1968), *Collin's Pocket Guide to Wildflowers*, London: Collins
Hubbard, C. E. (1984), *Grasses* (3rd edn), Harmondsworth: Pelican
Keeble Martin, W. (1982), *The New Concise British Flora in Colour*, London: Michael Joseph and Ebury Press
Phillips, R. (1977), *Wild Flowers of Britain*, London: Ward Lock and Pan Books
Phillips, R. (1978), *Trees in Britain, Europe and North America*, London: Ward Lock and Pan Books
Phillips, R. (1980), *Grasses, Ferns, Mosses and Lichens of Great Britain and Ireland*, London: Ward Lock and Pan Books
Rose, F. (1981), *A Wild Flower Key: British Isles and North West Europe*, London, Warne

Soils

Bridges, E. M. (1978), and reprints, *World soils* (2nd edn), London: Cambridge University Press (especially the associated transparency collection displaying soil profiles).
Briggs, D. J. (1977), *Sources and Methods in Geography: Soils*, London: Butterworth
Smith, R. J. and Atkinson, K. (1975), *Techniques in Pedology: a Handbook for Environmental and Resource Studies*, London: Elek Science
White, R. E. (1979), *Introduction to the Principles and Practice of Soil Science*, Oxford: Blackwell Scientific Publications

Index